Back to Basics in Physiology

Back to Basics in Physiology

O_2 and CO_2 in the Respiratory and Cardiovascular Systems

Juan Pablo Arroyo
Internal Medicine Resident
Tinsley R. Harrison Society Scholar
Vanderbilt University — School of Medicine

Adam J. Schweickert
Attending Physician
Hospitalist Medicine — Pediatric ICU
St. Barnabas Medical Center

AMSTERDAM • BOSTON • HEIDELBERG • LONDON
NEW YORK • OXFORD • PARIS • SAN DIEGO
SAN FRANCISCO • SINGAPORE • SYDNEY • TOKYO

Academic Press is an imprint of Elsevier

Academic Press is an imprint of Elsevier
125, London Wall, EC2Y 5AS
525 B Street, Suite 1800, San Diego, CA 92101-4495, USA
225 Wyman Street, Waltham, MA 02451, USA
The Boulevard, Langford Lane, Kidlington, Oxford OX5 1GB, UK

Notices
Knowledge and best practice in this field are constantly changing. As new research and
experience broaden our understanding, changes in research methods or professional practices,
may become necessary.

Practitioners and researchers must always rely on their own experience and knowledge in
evaluating and using any information or methods described herein. In using such information or
methods they should be mindful of their own safety and the safety of others, including parties for
whom they have a professional responsibility.

To the fullest extent of the law, neither the Publisher nor the authors, contributors, or editors,
assume any liability for any injury and/or damage to persons or property as a matter of products
liability, negligence or otherwise, or from any use or operation of any methods, products,
instructions, or ideas contained in the material herein.

ISBN: 978-0-12-801768-5

Library of Congress Cataloging-in-Publication Data
A catalog record for this book is available from the Library of Congress

British Library Cataloguing-in-Publication Data
A catalogue record for this book is available from the British Library

For Information on all Academic Press publications
visit our website at http://store.elsevier.com/

Working together
to grow libraries in
developing countries

ELSEVIER Book Aid International

www.elsevier.com • www.bookaid.org

DEDICATION

To our wives, Denise and Valentina, for their unwavering support of our every endeavor, both aimless and not so aimless.

CONTENTS

Acknowledgements ix
Preface... xi

Chapter 1 Cellular Respiration and Diffusion 1
Introduction..1
O_2 and CO_2 for One Cell: Mechanics of Single Cell Gas Exchange.....3
Review of the Physical Properties of Gases 7
Review of Diffusion and Gradients 10
Diffusion and the Cell 14
Development of Multicellular Organisms from Single Cells,
O_2 and CO_2 for Trillions of Cells 16
Clinical Vignettes..17

Chapter 2 Functional Anatomy of the Lungs and
Capillaries: Blueprints of Gas Exchange...........................**19**
Functional Anatomy of Gas Exchange 20
Functional Histology of the Capillaries and the
Alveolar−Capillary Unit 27
Clinical Vignettes..29

Chapter 3 Lung Mechanics: Putting the Blueprints of
Gas Exchange into Action...**31**
In and Out: How Gas Moves 33
Pleural Pressures: Negative versus Positive Pressure..........................41
Lungs Outside the Body: Tissue Dynamics 45
Clinical Vignettes..57

Chapter 4 The Respiratory Cycle **61**
What are Lung Volumes and Capacities?..61
Alveolar Ventilation and Dead Space Ventilation 65
Composition of Alveolar Air..69
Clinical Vignettes..77

Chapter 5 Gases Inside the Body, Liquid Transport...........................**81**
Getting into Blood, Henry's Law, and Why We
Need Red Blood Cells 81

Why are RBCs So Special?..83
Oh Marvelous Hemoglobin! 86
O_2 Content and O_2 Delivery...90
Dynamics of O_2–Hemoglobin Dissociation Curve:
How Does it Know Where to Deliver Its Cargo?..............94
Transport of CO_2 in the Blood.......................................101
Clinical Vignettes...103

Chapter 6 The Alveolar–Capillary Unit and V/Q Matching.............107
The Alveolar–Arterial Difference: How Good is the
Lung at Exchanging O_2 and CO_2?..................................107
Ventilation/Perfusion Relationships: Matching the
Movement of O_2, CO_2, and Blood 110
Clinical Vignettes...114

**Chapter 7 Regulation of O_2 and CO_2 in the
                    Body and Acid/Base 115**
What Actually Drives Ventilation?....................................115
CO_2, H_2O, Carbonic Acid, HCO_3^-, pH, and pKa—What? 118
Henderson–Hasselbalch Equation 120
Changes in CO_2 and pH: Respiratory Acid/Base Disorders.............122
Clinical Vignettes...124

**Chapter 8 Clinical Recognition: Signs and Symptoms of
                    Respiratory Distress and Their Physiologic Basis.............125**
Causes of Respiratory Failure 126
Ways that Ventilation Can Be Improved...........................131
Clinical Signs of Respiratory Failure................................132
Putting it All in Context 137

Chapter 9 Clinical Integration 139
Case #1...140
Case #2...144
Case #3...145
Case #4...149
Case #5...151
Case #6...153

Appendix: Respiratory Devices 155
Bibliography...163

ACKNOWLEDGEMENTS

We wish to thank Mara Conner, Jeffrey Rossetti, and the rest of the Elsevier staff for the time and hard work that went into helping to make this book a reality.

We also wish to thank all those who provided their insight and suggestions throughout the writing of this book, with a special thanks to Dr. Gary Kohn.

PREFACE

The whole idea for this series arose from the physiology classroom and hospital teaching rounds. We realized that both in the classroom and on the wards, students and residents had a fair amount of knowledge regarding individual organ systems. However, there was still room for improvement regarding how all the organ systems integrate in order to respond to a particular situation. This book series is an attempt to bridge the gap of knowledge that divides organ from body, and isolated action from integrated response.

Our goal is to create a series of books where the primary focus is the integration of concepts. The books in the series are written so that hopefully they are easy to read, and can be read from beginning to end.

It is our belief that if you truly understand something, you should be able to explain in a simple way. Therefore, we aim to tackle complicated topics with simple examples. And we hope that by the end of any book in this series, further more complex reading (e.g., the latest journal articles) should prove far easier to understand.

We hope you enjoy reading these books as much as we enjoyed writing them.

Other books in the series include:

Back to Basics in Physiology: Fluids in the Renal and Cardiovascular Systems (ISBN: 9780124071681)

Back to Basics in Physiology: Electrolytes and Nonelectrolyte Solutes in the Body (ISBN: 9780128017692)

Cellular Respiration and Diffusion

INTRODUCTION

Breathing in and out is key to staying alive. It's so important that even when we forget to breathe, our nervous system picks up the slack and keeps going. The process of breathing provides oxygen and removes carbon dioxide from the body. This process is essential to sustaining each and every cellular task within our bodies. The focus of this book is how the body achieves this seemingly simple process. We will take you from a single cell and how it regulates oxygen and carbon dioxide to the large-scale gas transport and delivery in the body under normal and pathologic conditions. So, sit back, relax, and take a deep breath!

If indeed you take a breath right now, you will breathe in air. Air in the atmosphere is a simply a mixture of gases. Atmospheric air, as it exists today, consists of about 21% oxygen, 78% nitrogen, 0.04% carbon dioxide, and some other miscellaneous gases such as argon. (Carbon dioxide makes up so little of the atmospheric air that it even gets beat out by argon, which weighs in at 1%. Seriously!)

But it wasn't always this way. In fact, over 2.5 billion years ago, things weren't looking too good for our oxygen-loving brethren. There was almost no oxygen in the atmosphere, and there was very little food around. So, some opportunistic little buggers called cyanobacteria took the warmth of the sun and made sustainable energy out it, much like plants do today. In the process they gave off oxygen as "waste."

Little by little cyanobacteria began filling up the oceans with oxygen. The dissolved oxygen began to diffuse throughout the water (hopefully you'll remember the principles of diffusion from our last book "Back to Basics in Physiology: Fluids in the Cardiovascular and Renal Systems"), and as the oceans filled with this "waste product" it diffused into the atmosphere. Over the next two billion years, the concentration of oxygen in the air reached the 21% we know and enjoy today.

Back to Basics in Physiology. DOI: http://dx.doi.org/10.1016/B978-0-12-801768-5.00001-0

As oxygen became more and more plentiful in the environment, creatures began using this oxygen to create energy from available food sources more efficiently, and were able to grow larger than their non-oxygen-consuming counterparts. With size came more food consumption and a greater need for mobility, and with mobility and size came more energy utilization. Over time, organisms migrated from the water to land. Cyanobacteria made room for plants in the sea and on land, which produced even more oxygen. As organisms developed ways to use this newfound energy (e.g., growing brains!), they developed a larger need for oxygen, produced more carbon dioxide, and along the way came up with some pretty ingenious mechanisms to ensure *constant* oxygen delivery and carbon dioxide removal.

In our bodies today, out of the millions of functions that need to be carried out minute by minute in order to allow for life to proceed "uneventfully," oxygen (O_2) and carbon dioxide (CO_2) exchange are arguably two of the most important processes our bodies require to stay alive. If the human body is deprived of oxygen, it will die far quicker than if deprived of food or water. If someone removed your kidneys right now, you would live for potentially several days. If they removed your heart or your lungs, the main organs responsible for moving the oxygen and carbon dioxide around the body, you would die within minutes. In fact, doctors' primary goals in the setting of any medical emergency always revolve around bringing back or "stabilizing" a patient's oxygen delivery, and to a lesser extent, carbon dioxide clearance. In fact, the classic ABCs of patient care (what doctors need to worry about first!) stand for Airway, Breathing, and Circulation. But why exactly are these two items so important?

O_2 is consumed and CO_2 is produced by all living cells in the body every second of every day in a process called aerobic cellular respiration. This process is absolutely vital to creating the energy that keeps the cells alive. O_2 and CO_2 allow for the most efficient energy extraction from the food we eat. In order to keep creating energy, these cells need a system that will move new O_2 in and take CO_2 out. So, before we go on to understand exactly how O_2 and CO_2 move in and out of the body, we need to take a step "in" and first understand why O_2 and CO_2 are important, and how they help create energy at the cellular level. Then we can move on to how these vital gases get in and out of cells and why blood is specialized to help aid this process. In the subsequent chapters, we will apply these concepts to the lungs and the rest of the cardiovascular system. By understanding how O_2 and CO_2 are used and how they move, the form and function of the rest of the pulmonary and cardiovascular systems will make sense intuitively.

Key

O_2 is consumed and CO_2 is produced in the creation of energy.

O_2 AND CO_2 FOR ONE CELL: MECHANICS OF SINGLE CELL GAS EXCHANGE

A cell is the most basic unit of life (ignoring viruses, which are a bit of a gray area). As such, it needs to be able to grow and respond to threats in its environment long enough to reproduce before eventually dying. Biochemically speaking, this involves a myriad of complex tasks. However, in order to perform all of these incredibly complex tasks, one thing is key: energy! Energy is needed for every major process the cell undertakes: movement of ions, signaling, and reproduction. We need energy for everything. But where does this energy come from?

Role of Oxygen (O_2) and ATP

Much like how money is used to allow us to survive in a modern economy, cells must have a form of "energy currency" that allows them to rapidly generate and store energy that can be used at a moment's notice. In organisms, this energy is most commonly stored as ATP, or adenosine triphosphate. Adenosine is a nucleoside. Nucleosides (a nitrogenous base with a carbohydrate backbone) are some of the most ubiquitous chemical compounds found in life. They are the building blocks of DNA and RNA, so your body has loads of them on hand. If multiple phosphate molecules are added to them, they become increasingly energy rich. *In short, it is energy in the form of ATP that fuels life.* As we shall soon see, oxygen makes ATP formation a heck of a lot more efficient. And efficient is good!

Generally speaking, ATP can be made without the help of oxygen. Many microorganisms from many walks of life live in some of the most hostile and oxygen-poor environments on this earth, but they can still thrive. They need to worry about providing fuel for only *one* little cell, though. The human body, on the other hand, is made up of *trillions* of cells, and within it ATP is broken down and formed and broken down and formed over and over again, millions of times a day. This pathway is so active that the body effectively turns over its own body weight in ATP every day! You can imagine then that ATP production can become exceedingly expensive to produce. Thankfully, oxygen helps us make ATP creation a lot easier.

Let's look at ATP fabrication and recycling a little bit more closely, shall we? As we just mentioned, oxygen allows for the efficient creation of energy in the form of ATP. In more general terms, energy is extracted from the food we eat. As such one of the key molecules in all the food we eat is glucose. The process through which oxygen is used to extract energy to make ATP from glucose is called cellular respiration (Figure 1.1):

$$\text{Glucose} + O_2 \rightarrow CO_2 + H_2O + ATP$$

Key

O_2 is consumed and CO_2 is produced during aerobic respiration. The product is energy!

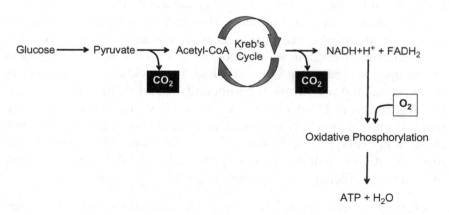

Figure 1.1 Aerobic cellular respiration is the process through which cells use glucose and oxygen to produce ATP and H_2O, with CO_2 as a byproduct of the biochemical reactions.

Clinical Correlate

Ischemia
Ischemia is what happens when cells suddenly are unable to receive oxygen and get rid of carbon dioxide. Specifically, the term is used to describe a loss of blood flow. As we'll see in later chapters, one of the main functions of blood is to deliver O_2 to tissues and remove carbon dioxide. When there is no blood flow, there is no O_2 delivery, and there is no CO_2 removal. Therefore, cells are no longer able to produce energy, and they begin to malfunction. One of the best examples of this is myocardial ischemia—a heart attack. When blood flow to a portion of the heart muscle stops, the heart muscle cells can't make energy. This causes inflammation and abnormal functioning of these cells. Common clinical manifestations of a myocardial infarction are pain and arrhythmias arising from the infarcted tissue.

Role of Carbon Dioxide (CO_2)

The amount of CO_2 that is in the air we breathe is relatively low, but inside the body the amount of CO_2 is much, much higher. As O_2 is actively being consumed during cellular respiration, CO_2 is being produced as a byproduct of the same biochemical pathway (Figure 1.1). Remember: While O_2 is being consumed, CO_2 is being produced. Similar to what happens with O_2, the production of CO_2 by the cell is closely linked to metabolism; the higher the metabolic rate, the more CO_2 produced. The major goal of metabolizing food is to break down the food into its simplest chemical form (usually glucose) and then to remove hydrogen ions and electrons from it. The removal of hydrogen ions and electrons will ultimately power an enzyme called ATP synthase. This enzyme creates ATP, and in doing so creates usable energy. There are many biochemical reactions involving the removal of hydrogen ions and electrons from food, and they differ depending on whether the food is a sugar, a protein, or a fat. Some of these reactions, called decarboxylation reactions, result in the removal of a carbon atom, and it is from these reactions that CO_2 is generated.

CO_2 is not a useless byproduct of metabolism though; it has an extremely important role in the body as an acid base buffer, as we will see in further chapters. For now, suffice it to say that any excess accumulation of CO_2 within the cell is unwanted and could disrupt adequate cellular functioning; therefore, CO_2 must be continuously shuttled outside of the cell.

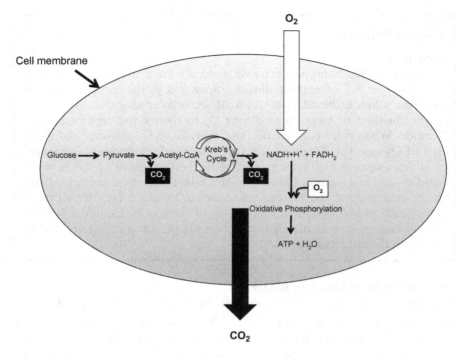

Figure 1.2 In a single cell, O_2 moves from the outside of the cell to the inside (white arrow), while CO_2 moves from the inside to the outside of the cell (black arrow).

Single Cell Exchange Requirements

We've established that during the aerobic production of ATP, O_2 is consumed and CO_2 is produced by the cell (Figure 1.2). Because O_2 gets consumed by the cell, it must first be brought to the cell from the outside, while CO_2 is produced and must be shuttled from inside to outside in order to prevent a toxic accumulation of CO_2 inside the cell. But how do these gases move across the cell membrane? Unlike ions, which require proteins to be shuttled in and out of the cell due to a lack of permeability, gases can freely diffuse in and out of the cell. Because gases are freely diffusible, the only thing that regulates the movement of O_2 and CO_2 across the membrane is the pressure difference between both sides of a membrane and the solubility of the gas. Therefore in order to fully understand the movement of gases between cells and in the body, a brief review of the basic principles regulating the behavior of gases in the environment is warranted.

REVIEW OF THE PHYSICAL PROPERTIES OF GASES

The physical and chemical properties that guide the diffusion of gases are far too complex to be entirely reviewed in this book. However, we will highlight the bare minimum that we believe is essential to understanding the movement of O_2 and CO_2 in the body. With that in mind, let us push forward!

There are four fundamental states of matter: solid, liquid, gas, and plasma. A simple way to define the differences in the states of matter is to think of the kinetic energy of their molecules. All molecules move constantly, and this movement has the capacity to do work. Kinetic energy is the energy that that these molecules possess due to the movement of their molecules. The more kinetic energy, the more they're going to move. Solids have the least amount of kinetic energy and plasma has the highest amount of kinetic energy. As the kinetic energy increases, molecule movement increases. Sugar-laden 4-year-olds running wild at a birthday party = high kinetic energy; the same 4-year-olds asleep after the sugar crash = low kinetic energy. As kinetic energy increases in the molecules that make up a given compound, it becomes harder for the compound to keep its shape as the intermolecular bonds weaken from all the motion. The more kinetic energy the molecules have, the more space they will occupy and the less likely they are to interact. Solids are solids because of the stable interactions between molecules. Gases have a much higher amount of kinetic energy; this means that gas molecules moving around all over the place take up a lot more space. (Keeping with our young child analogy, a sleeping child equaling low kinetic energy does not occupy that much space. A sugar-crazed toddler running around the house can feel as if no place is big enough to contain him or her.) Thinking of the matter in this way (and specifically, gases) leads us to the following point. There are four basic physical properties that significantly impact the behavior of gases by impacting their molecular kinetic energy in a manner of speaking:

- Number of particles
- Temperature
- Volume
- Pressure

Key

The key determinants of the behavior of a gas are the number of particles, its temperature, its volume, and its pressure.

Of these factors let's take a closer look at pressure, since this will become relevant when we discuss gas movement in the body. What exactly is pressure? Pressure is the amount of force that is applied by a particular compound in a given area. If that compound were a gas, it would be the force from all those collective collisions banging up against the sides of, say, a container holding said gas:

$$Pressure = \frac{Force}{Area}$$

Pressure is therefore a function of the strength between the collisions of the molecules in the gas and the amount of space these molecules have to move around in. So how exactly do we quantify pressure? There are various units that can be used: atmospheres (ATM), Pascals, pounds per square inch (PSI), Torr, among others. We will be using two particular units: millimeters of mercury (mmHg) and centimeters of water (cmH$_2$O). Both of these methods work in a similar fashion (Figure 1.3). A graduated glass column is filled with either mercury (Hg) or water (H$_2$O), and it's connected through an adaptor to wherever you want to

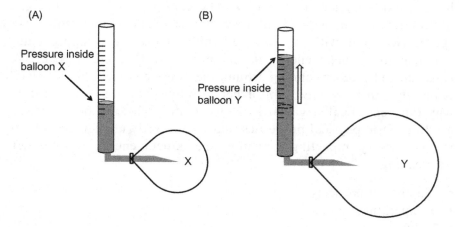

Figure 1.3 Displacement of a column of fluid (grey) allows for the measurement of pressures. (A) Low pressures only make the column of fluid rise slightly. (B) Higher pressure makes the column of fluid rise higher. The pressure, in either mmHg or cmH$_2$O, is the total amount of displacement measured in the column.

measure the pressure. The pressure inside the place of interest will displace the water or mercury a specific distance either up (high pressure) or down (low pressure). In the case of mercury this is measured in millimeters and in the case of water it is measured in centimeters. The amount of fluid that ends up getting displaced is measured. It's typically much easier to see a liquid than a gas, so this form of measurement has been historically convenient. Given that mercury has a greater density than water, mmHg are used for higher pressures (it is harder to displace a dense liquid so we need higher pressures), and cmH_2O are used for lower pressures (easier to displace a less dense liquid with a lower pressure). So whenever we mention mmHg or cmH_2O, what we are referring to is how much pressure is in a particular space. There are several pressures that we are required to memorize: first is the atmospheric pressure at sea level. This is the standard pressure of air at sea level, which is 760 mmHg. All further calculations in the book will be based on sea level atmospheric pressure!

Another concept that requires a brief mention (we'll touch on it again in Chapter 2) is that of partial pressures. We mentioned that the pressure of air at sea level is 760 mmHg. But air is simply a collection of gases! If we're thinking of these molecules as individuals, each with their own weight (oxygen, e.g., is heavier than nitrogen) and their own size (nitrogen, due to different electron configuration is actually larger than oxygen), then we can imagine that each individual gas collectively exerts its own pressure within the air! Thus, a partial pressure is simply the amount of pressure that an individual gas within a mixture exerts. For example, we said earlier that O_2 makes up 21% of atmospheric air, but this tells us the relative amount. Without knowing the total pressure, this number doesn't help us exactly. We want to know the amount of oxygen in absolute terms (e.g., its partial pressure in mmHg). At sea level, where we know that the total air pressure of the atmosphere is 760 mmHg, we can determine that 21% of this is 160 mmHg. This would be the value of the partial pressure of oxygen within the atmosphere at sea level. If we were to hypothetically increase the percentage of oxygen to 40%, but keep the total atmospheric air pressure the same, the partial pressure of O_2 would increase from 160 mmHg to 304 mmHg. Conversely, if we were to move much higher up away from sea level, where there is less gravitational force acting on molecules and a lower total air pressure (let's say 500 mmHg instead of 760 mmHg), the fraction of inspired O_2 (FiO_2) will remain the same at 21%, but the ABSOLUTE pressure of O_2 will decrease from 160 mmHg to 105 mmHg. Therefore it is important to consider both the total pressure and the fractional percentage that each gas we're studying represents.

REVIEW OF DIFFUSION AND GRADIENTS

In its simplest terms, diffusion is the movement of substance X from an area where there is a lot of X to an area where there is not that much X. When discussing gases, we can talk about a gradient from an area of high pressure to an area of low pressure along the pressure gradient (Figure 1.4). In our previous book, *Back to Basics in Physiology: Fluids in the Cardiovascular and Renal Systems*, we had defined diffusion as the movement of substance X using the term concentration rather than pressure. This was the case because we were talking mainly about solutes and solvents. Since now we are referring to gases we talk in terms of pressure. Other than pressure, the factors that modify the diffusion across a semipermeable membrane of any one substance in particular can be summarized with the following formula:

$$\text{Diffusion } \alpha \frac{\Delta P \times SA \times sol}{dist \times \sqrt{MW}}$$

where:

$\Delta P = (P_1 - P_2)$. The difference in pressures between compartment 1 (P_1) and compartment 2 (P_2). As you can see this is in the numerator; thus, the greater the pressure difference the greater the diffusion that will take place.

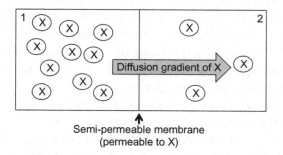

Figure 1.4 Diffusion of X from compartment 1 to compartment 2 follows the gradient that exists between A (high pressure) and B (low pressure).

SA = Surface area. How much membrane space is available for difussion to occur. Again, numerator. The more membrane through which exchange can occur, the more diffusion will take place.

sol = Solubility. Determined by two things: (1) the semipermeable membrane (e.g., if something is not soluble to the membrane it will never diffuse across no matter the pressure difference or surface area) and (2) the states of matter on either side of the membrane (e.g., is it diffusing from gas to gas? Liquid to gas? Gas to liquid?). We will approach this particular concept again in the upcoming chapters, but for now let us cover the highlights. When diffusion of gases is occurring solely as gases and does not involve liquids, the only impediment to diffusion will be how permeable the membrane is to a particular gas (Figure 1.5A). In contrast, when a gas is diffusing from a gas to a liquid (Figure 1.5B), conditions change. The gas must first dissolve in the liquid before it can diffuse throughout the liquid. This is when solubility becomes even more important, because how readily a gas will dissolve (e.g., in water) will have a large impact on its rate of diffusion. (This explanation applies to any fluid, but considering that water will be the basis of our discussions, we will continue to discuss solubility of gases in water.)

dist = Distance. How much distance is there from one compartment to another? As the distance increases, diffusion decreases (in this case this variable is in the denominator). This is especially

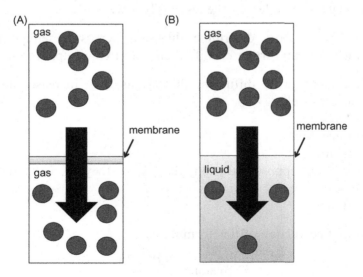

Figure 1.5 The solubility of a gas in a particular liquid will determine its diffusion into and through the liquid.

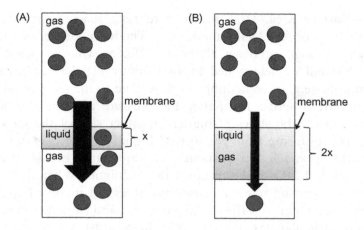

Figure 1.6 Distance is one of the key determinants of diffusion. As distance increases (gray area), diffusion decreases (black arrows). The converse is also true, as the distance decreases, diffusion increases.

important in certain clinical conditions, which we will approach later in the book. But for now, consider Figure 1.5; the liquid membrane dividing both compartments has a predefined distance (x) (Figure 1.6A), and as this distance doubles (Figure 1.6B), diffusion decreases. If distance were to decrease, diffusion would increase proportionally.

MW = Molecular weight. MW of the substance we're analyzing. Stated differently, how big is the molecule that is going to diffuse? The bigger the molecule, the less easily it will diffuse.

These five factors determine diffusion across a semipermeable membrane. We can further classify them into two groups (Figure 1.7):

1. Factors that favor diffusion; that is, as they increase, diffusion increases as well:
   - ΔP
   - Surface Area
   - Solubility
2. Factors that oppose diffusion; that is, as they increase, diffusion decreases:
   - Distance

A simplified version of this formula is:

$$\text{Diffusion} \propto \frac{\Delta P \times SA}{dist}$$

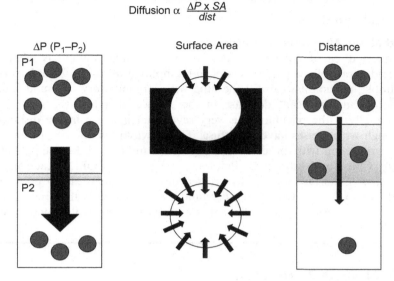

$$\text{Diffusion} \; \alpha \; \frac{\Delta P \times SA}{dist}$$

Figure 1.7 The formula for diffusion has two nonvariable factors: solubility (sol) and MW (molecular weight). These are intrinsic to each substance and can't be readily altered. From the remaining three variables, two favor diffusion (ΔP, and surface area) and one opposes diffusion (distance). From this image we can see that as ΔP or surface area increases, diffusion (black arrows) increases. However if distance increases, diffusion decreases.

In this simplified version, we have eliminated solubility and molecular weight. This is because solubility can't be easily altered. In fact, in the body the solubility of different gases is relatively fixed. Therefore it's not going to be a variable that affects diffusion in a significant way under steady state conditions (although we will see that it does explain some of the differences we see later between oxygen and carbon dioxide!). Additionally for simplification we will not include MW since it isn't exactly something we can modify.

Key

As ΔP and surface area increase, diffusion increases. As distance increases, diffusion decreases.

This formula explains diffusion throughout the entire body and for every system. This means that if you understand this formula you will be able to understand the pathophysiological mechanism of diffusion problems everywhere in the body.

Clinical Correlate

Pathologic Alterations in Diffusion

In many diseases that impair adequate gas exchange, ultimately what is altered is one of the parameters of our simplified diffusion formula. For example, pneumonia can fill the lung with inflammatory cells and fluid can have a significant decrease in the surface area and the ΔP, and potentially if the pneumonia is very severe, an increase in the distance through which gases have to diffuse! This is why patients with pneumonia can present with difficulty exchanging O_2 and CO_2. Likewise patients with Chronic Obstructive Pulmonary Disease (COPD) can have decreases in surface area and increased distance, which lead to poor O_2 and CO_2 exchange. Once we understand the root cause of the disease, we can try to orient our treatment to reestablish normal function of the lungs.

DIFFUSION AND THE CELL

After our brief review of diffusion, let's go back to discussing our single cell example from the previous paragraphs. As we mentioned previously, cells in the body consume O_2 and produce CO_2. This is happening in each and every cell. Each single cell is consuming O_2 and therefore, O_2 must diffuse through the cell wall from the outside in. Along the same lines, CO_2 is being produced and it must diffuse from the inside of the cell to the outside. Consider our simplified diffusion formula:

$$\text{Diffusion } \alpha \frac{\Delta P \times SA}{dist}$$

If we're talking about a single cell, surface area is going to be more than adequate to allow for both diffusion of O_2 and CO_2 and the distance; that is, the width of a single cell membrane is not a huge obstacle to diffusion, therefore the most important factor determining the diffusion of O_2 and CO_2 in this example is the ΔP (Figure 1.8).

For O_2:

- $\Delta P = (P_1 - P_2)$. P_1 is outside the cell, and P_2 is inside the cell. Given that O_2 is being consumed inside the cell, we can automatically assume that the pressure of O_2 inside the cell is going to be lower than the pressure of O_2 outside. Therefore, if P_2 is less than P_1, an

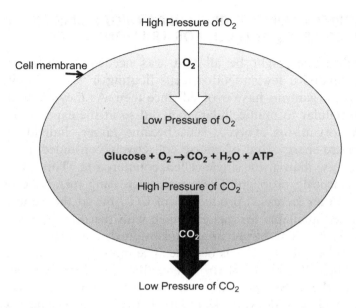

Figure 1.8 The diffusion of O_2 and CO_2 across the cell membrane follows their individual pressure gradients (see arrows).

O_2 gradient is generated from the outside of the cell toward the inside. This means that O_2 will tend to diffuse toward the inside of the cell.

For CO_2:

- $\Delta P = (P_1 - P_2)$. P_1 is inside the cell, and P_2 is outside the cell. The situation is reversed in this case. CO_2 is being produced; therefore, we can automatically assume that the pressure of CO_2 inside the cell is higher than outside the cell. This generates a favorable CO_2 diffusion gradient from the inside to the outside of the cell.

If we take all of these considerations into account, in the case of the single cell we can say that the requirements of O_2 consumption and CO_2 excretion are easily met. However, more complex organisms are made up of an increasing number of cells, and as the number of cells increased, so did the metabolic requirements. Unfortunately, what did not increase was the surface area! That meant organisms had to develop a way of increasing the surface area that's available for exchange while maintaining structural integrity.

DEVELOPMENT OF MULTICELLULAR ORGANISMS FROM SINGLE CELLS, O_2 AND CO_2 FOR TRILLIONS OF CELLS

Simple diffusion might be all that was needed if organisms never evolved beyond a few individual cells floating in a sea. However, as you know, organisms have evolved since then. As they made the jump from unicellular to multicellular, they did so at the expense of surface area! As organisms evolved, cells became larger, individually consumed more energy, and of course collectively consumed a great deal more energy than their single-celled counterparts. This presented a *huge* engineering problem! There were increasing metabolic demands for O_2, and an increased production of CO_2, but at the same time the surface area available for gas exchange with the outside environment was decreasing. The bigger an organism grew, the farther away from the atmosphere its inner cells became. The more the distance increased between the cells and the environment, the more difficult it became to oxygenate them. So, how do you feed all those hungry cells with oxygen? And how do you clear out all the waste products when so many cells are so far away from their external environment? Thankfully, nature developed a solution: specialized labor.

Any student of nature can guess that there are probably some advantages to being multicellular from an evolutionary standpoint, no? The most obvious one that comes to mind is specialization of labor. If you're a single cell, you have to be a jack-of-all-trades and a master of none. But if you're a multicellular organism, then you can have cells that specialize and spend their entire existence devoted to just one task. You can create cells that specialize in sensing your environment (eyes, ears, nose, tongue); cells that protect you from said environment (skin, immune system); cells that help you get around your environment with locomotion (skeleton, muscles); cells that absorb food (digestive system); cells that keep all these systems organized and communicating together (nervous system, endocrine system). But most vitally important (pun intended) would be the cells whose job is to make sure that all of these specialized cells have enough energy to do all of these things; cells that help supply the rest of the cells in our body with O_2 and help rid the body of excess CO_2. And these are the cells that make up the respiratory system, cardiovascular system, and blood.

In the next chapter, we're going to look at how the body engineered a solution to this problem: how to deliver oxygen from the atmosphere to the cell, and how to deliver carbon dioxide from the cell to the atmosphere. As we'll see, rather than trying to reinvent the wheel, the body relies on diffusion to do most of the work. At every level of the body, diffusion is what drives gas exchange. Whether it's through creating a larger gradient, maximizing surface area, or minimizing the distance oxygen and carbon dioxide need to travel, the body engineered a system where diffusion does most of the work. It will become apparent in subsequent chapters that the lungs, heart, blood vessels, and blood all work in symphony to make sure that a healthy gradient from the atmosphere to the cell is always maintained. They make sure that there is as much surface area as possible through which oxygen and carbon dioxide can diffuse, and they make sure that there is as little a distance over which it needs to take place. They simply serve to get the gases *close* to where they need to go, and they let diffusion do the rest. This is a recurrent theme throughout the body, and thus will be a recurrent theme throughout this book.

Key

Diffusion is what drives gas exchange to the trillions of cells within the body. Fresh gradients, large surface area, and short distance is how the body keeps oxygen flowing in and carbon dioxide flowing out.

CLINICAL VIGNETTES

A 56-year-old stockbroker with a 20-pack/year history of smoking and untreated hypertension presents to the emergency department complaining of 10/10 chest pain that started suddenly approximately 45 minutes ago while he was at work. The pain is "crushing" and radiates to his left arm. He is short of breath, and feels light-headed. An EKG is performed, which shows a clear S-T segment elevation in leads V4−V6, consistent with the diagnosis of left heart wall myocardial infarction (a heart attack).

1. Why does this patient have pain?
   A. Blood is building up in his lungs secondary to his decreased lung function.
   B. The lack of O_2 leads to inflammation, pain, and eventually cell death.
   C. He has a bad case of gastroesophageal reflux disease.

Answer: B. This patient is clearly having a left wall myocardial infarction. The lack of O_2 delivery to the cardiac muscle leads to ischemia and subsequent inflammation of the O_2-deprived cells. With inflammation comes pain. If the area of infarction is large enough and decreases the ability of the heart to pump out blood, some fluid could be potentially building up in his lungs. This however would not present as pain, but rather as dyspnea (i.e., difficult or uncomfortable breathing). Although gastroesophageal reflux can also cause chest pain, this patient's presentation is more likely to be secondary to a myocardial infarction.

CHAPTER 2

Functional Anatomy of the Lungs and Capillaries: Blueprints of Gas Exchange

The overall goal of the lungs is to provide the exchange of O_2 and CO_2 in order to sustain life. Although this may sound relatively easy (O_2 in, CO_2 out), the means by which this is accomplished for trillions of oxygen-hungry, carbon dioxide-wasting cells is incredibly complex. We know that the lungs want to let diffusion do most of the work, so they've evolved as organs. They've come up with some incredible space-saving tricks to create a huge surface area for gas exchange packed into a relatively tiny space. This surface area is so large that the lungs of an average adult male, when spread out flat on the ground, would take up about 70 to 80 m^2, which is about half the size of a singles tennis court! The lungs, however, don't work alone. Have you heard of capillaries? They're the alveoli's partners in crime, so to speak. They are an extensive network of very small blood vessels (only one-cell thick!), and serve as the interface between the blood and all the organs in the body, including the lungs. Capillaries allow for diffusion to take place between blood and peripheral tissues, as well as between blood and lungs. If you thought that the lungs had a huge surface area available for exchange, it's nothing compared to the capillaries! If you pieced together every single capillary and arranged them end to end, it's been estimated they would measure about 60,000 miles in length! Their combined surface area is between 800 and 1000 m^2; that's an area of about three tennis courts!

If you think about the basic needs of the body (lots of oxygen-hungry cells producing carbon dioxide at a very high rate), and you understand the physical principals governing diffusion and gases, it will become less important that you understand every detail of the anatomy (unless you're planning on learning how to cut into it one day). Instead, it's more important that you understand how the anatomical form allows for the desired function because the two, form and function, are intimately linked. So rather than asking you to memorize the name of every last artery, vein, and alveolus, our goal in this

Back to Basics in Physiology. DOI: http://dx.doi.org/10.1016/B978-0-12-801768-5.00002-2

chapter will be that you understand the basic blueprints of gas exchange. Once you understand how the body decided to engineer a solution to the problem, we'll go into how the blueprint works, the mechanics of gas exchange.

FUNCTIONAL ANATOMY OF GAS EXCHANGE

In order to understand how diffusion takes place in the ventilation system, it's important to have a basic understanding of the ventilation system's anatomy. So let's explore the actual tissues, organs, and systems that the body uses. As such, we will be dividing our discussion into two sections:

1. **The lungs and alveoli.** As we mentioned earlier, the lungs are the quintessential organs of ventilation. The lungs are the place where O_2 and CO_2 are exchanged with the atmosphere. This happens through an exchange membrane made up by the alveoli. The alveoli are the terminal divisions of the airways and are the functional units of the lungs. They are air-filled sacs that have a one-cell thick wall, which allows for the rapid exchange of O_2 and CO_2 between alveolar air and the blood in the pulmonary capillaries. They are the "gas side" of our exchange membrane.
2. **Blood and capillaries.** The capillaries are the liquid side of this exchange membrane. They are the contact point through which this O_2 and CO_2 diffuse, both at the level of the lungs and at the level of every organ in the body. They are also only one-cell thick. At the level of the lungs, there is such a dense network of capillaries intimately linked with each alveolus that they have been described together as the alveolar–capillary unit. Here on earth, oxygen and carbon dioxide "prefer" to be in their gas state. They have a fairly low solubility in liquid. Blood, as we'll learn in much greater detail in Chapter 5, is specialized in carrying these two gases in the liquid phase. Because this whole system relies on diffusion, because these gases are relatively less soluble in the liquid phase, and because the capillaries are responsible for feeding every organ in the body, the capillaries need to cover an even larger amount of surface area than the lungs.

Functional Anatomy of the Lungs

The lungs developed as a way to increase the gas exchange surface area, so that the entire body's requirements of O_2 input and CO_2

output are met. In other words, the O_2 and CO_2 that the lungs exchange is the *combined total* of O_2 being consumed and CO_2 produced by each individual cell in the body! The way in which the lungs achieve this is through an exponential division of its branches until they get so thin and in such close contact with the capillaries that diffusion of O_2 and CO_2 can take place through a large surface area!

A simple way to understand the anatomy of the lungs is to think of a large tree. There is only one trunk for most trees, and this trunk progressively divides thousands of times until the smallest branches give off leaves (which actually carry out the gas exchange with the atmosphere!). The leaves increase the surface area that is available for exchange, and each tree has millions of leaves relative to the single trunk. The lungs work exactly like this. They're a series of tubes (similar to the trunk and branches) that give off sequential segments (23 to be exact) until they form the terminal air sacs or alveoli (similar to the leaves) where the exchange takes place. In an average adult human lung, there are estimated to be some 480 million alveoli! Just like what happens in the trees, we can divide the lungs into segments that *can't* carry out gas exchange (trunk and branches), and places that *can* carry out gas exchange (leaves). Places that *can't* carry out gas exchange are known as conducting airways, and places that *can* carry out gas exchange are known as exchange airways (Figure 2.1):

1. **Conducting airways**. These airways guide air movement in and out of the body, but exchange *does not* take place through them. They are individually broken down into (in the order of larger conducting airways to smaller) the trachea, bronchi, bronchioles, and terminal bronchioles. They comprise the initial 15 to 16 divisions of the airways.
2. **Exchange airways**. Diffusion of O_2 and CO_2 takes place through the exchange airways. They comprise from the 17th to the 23rd division of the bronchial tree. The major site for gas exchange is at the level of the alveolar sacs (the 23rd division), however alveoli begin to appear lining the airways from the 17th division onward, therefore exchange can take place at anyone of these levels. Alveoli are the functional units of the lung. They are where the majority of gas exchange takes place.

These two major subdivisions have important consequences with regard to airflow and gas exchange, as we will see in the upcoming

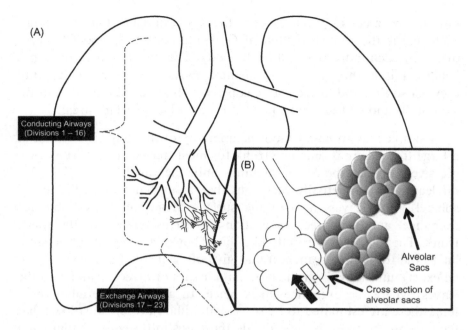

Figure 2.1 Basic anatomy of the lungs. (A) The lungs with a schematic representation of the branching airways. (B) Exchange airways including alveolar sacs (where gas exchange takes place).

chapters. For now let us leave you with the following important concept differentiating conducting and exchange airways. The areas of the lung where gas exchange *does not* occur are called *dead space*, and they will *not* take a part in diffusion. This is an important concept because even if air is moving in and out of the lungs, if there is a large volume of dead space (area of the lung that is not exchanging O_2 and CO_2), diffusion of O_2 and CO_2 will be impaired.

Functional Histology of the Lungs

Given the diverse functions of the airways, the cells and tissue that make up the conducting and exchange airways are inherently different. The overall goal of the airways, especially the conducting airways, is to get air to the alveoli. In order to do that, these airways must be able to deal with a couple of issues:

• Air is filled with microparticles and pollutants that can damage the alveoli. This means that the body must find a way to filter out these particles as much as possible so that alveoli are not damaged.

- Viruses and bacteria must be eliminated so that the airway is kept infection free.
- Even though there will be varying pressures (positive and negative) within the airways they must always be kept open to allow for adequate air flow in and out of the lungs.

> **Key**
>
> The ultimate goal of the airways is to keep clean, fresh air being supplied to the alveoli.

Keeping all of these characteristics in mind, let's take a look at how the airways are made up (Figures 2.2 and 2.3). For simplicity purposes we will divide the airways into three segments:

- Large conduction airways
- Small conduction airways
- Alveoli

Each of these segments is represented in Figures 2.2 and 2.3. The "layers" of tissue that are found in each segment are directly related to their function. Take a look at the large conduction airways in Figure 2.2. As the name implies, large conduction airways help conduct clean atmospheric air to the exchange membrane. As such, they have to

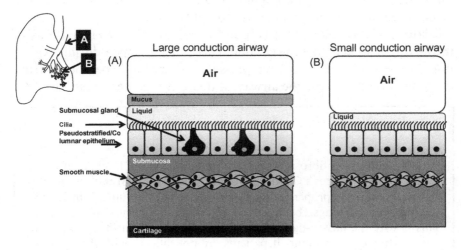

Figure 2.2 Histologic structure of the different types of airways. (A) Large conduction airways are the only airways that have cartilage, whereas (B) small conduction airways have smooth muscle but no cartilage.

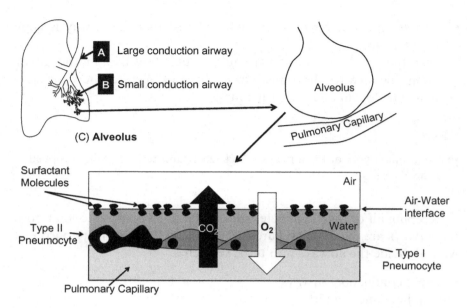

Figure 2.3 Histologic structure of the alveolus. The actual alveolar membrane is made up by Type I pneumocytes. The Type II pneumocytes are the "stem cells" of the alveoli. They can generate more Type I Pneumocytes in case of injury. The air–water interface, where diffusion beings, is lined with surfactant molecules, which help decrease the surface tension. Type II pneumocytes are in charge of surfactant production.

clear the air of small particles along with microorganisms. A thin layer of mucus helps clear the airway of small debris and microorganisms. This mucus layer is constantly being propelled toward the pharynx by tiny little mobile hairs (cilia) lining the airways that allow the mucus to be coughed up and spit up or swallowed! (This is called the mucociliary escalator). The cartilage that is found in the first 10 to 11 divisions of the airway help keep the lower conduction airways open during expiration. As we will see later on, the pressure required to expel the air from the lungs would collapse the airways if this cartilage were missing!

Clinical Correlate

Cystic Fibrosis

Patients with cystic fibrosis (CF) have mutations in a Cl^- channel called CFTR (Cystic Fibrosis Transmembrane Regulator). Mutations in this channel inhibit the proper production of mucus in the lungs, which leads to thick mucus plugs, bacterial overgrowth, and recurrent infections. This is why the day-to-day care of respiratory function in patients with CF is extremely important. If patients are not diligent with their various treatments aimed at maintaining good airway hygiene (clearance of mucus and secretions from the airway), they are likely to develop bacterial overgrowth and infection. In addition to providing these patients with antibiotics aimed at the unique types of bacterial infections they get due to their abnormal clearance mechanisms, in the event these patients *do* develop signs of respiratory difficulty and require hospitalization, much of their medical plan involves "tuning up" their mechanics vis-à-vis aggressive airway hygiene therapies.

As the airways continue to branch and become progressively smaller, the cartilage, the mucus, the smooth muscle layer, and the pseudostratified epithelium disappear giving way to the alveolar membrane. The idea is to get the most amount of debris- and microbe-free air down to the alveolar membrane so exchange of O_2 and CO_2 can take place undeterred and in an environment that is as clean as possible.

Clinical Correlate

Bronchiolitis

Anyone who has done a Pediatrics rotation in winter (or knows a baby) has probably encountered the all too common clinical condition known as bronchiolitis. Most notoriously caused by the Respiratory Syncytial Virus (RSV), it is a condition characterized by viral infection resulting in inflammation of the lower airways (bronchioles). This inflammation in infants often results in cell death, sloughing off of dead cells, mucus production, and swelling in the lower airways. As the cells die so do the cilia. While adults' airways are larger and also have small accessory connections between the lower lung segments, infants' airways are very small and lack these connections. Because the mucociliary escalator is impaired, and these children have smaller airways to begin with, very little can be done medically to help these children beyond deep nasal suctioning and oxygen. The body simply needs to wait until the new airway cells and cilia can regrow! The airway cells can regrow in a matter of 3 to 4 days, but cilia can take up to 2 weeks to regrow. For this reason, many of the children who are sick enough to require hospitalization often have more than one virus at a given time, with the second infection starting before sufficient time was allowed for recovery from the first.

As we mentioned previously, the functional unit of the lung is called the alveolus (alveoli is the plural form of the noun). The alveolus is the functional unit because this is where the exchange of O_2 and CO_2 with the blood takes place. As we were explaining earlier, we need a common exchange membrane in order to supply the needs of O_2 intake and CO_2 clearance for *all* the cells in the body. In order for this to work, this exchange membrane needs to be in constant contact with the atmospheric air. Their collectively large surface area and the thinness of their cellular wall are what make their histology so important. If, for example, we tried to just breathe through our skin (which is in constant contact with the air), we'd have a total body surface area of approximately 1.7 m^2 to use as an exchange membrane. The alveolar membrane within the lungs provides more than 20 times more surface area, at 75 m^2! It also provides a very small distance over which the gas has to travel. Once the gas is at the level of the alveoli, it has to cross the alveolar membrane to make its way into the blood. As we mentioned in the beginning of this chapter, this membrane is only a single cell-layer thick. If we were to compare this to our skin, even if we only needed the gas to diffuse through the *outermost* layer of skin (in reality there are several more), we'd be looking at a distance about 1000 times greater for the gas to travel ($\sim$1 millimeter vs. 1 micrometer)!

The cells that make up the alveolar membrane are called pneumocytes (Figure 2.3), and they can be subdivided into two types:

- **Type I Pneumocytes**. These are the cells that actually make up the exchange membrane. They form the outer wall of the alveoli and are in direct contact with the interstitial space and the pulmonary capillaries. It is through these cells that gases must diffuse in order to reach the blood.
- **Type II Pneumocytes**. These cells are in charge of generating more type I pneumocytes when they are damaged and need replacement. They are also in charge of making a substance rich in lipids called surfactant, which makes the expansion of the lungs a lot easier! (We'll get into this a little later in the book.)

Clinical Correlate

Emphysema

Emphysema is a disease you have probably heard of. What you may not have known is that it is a disease that primarily screws up this simple diffusion system. It takes one of the collective alveolar membrane's greatest advantages, surface area, and turns it against itself. Whether due to chronic cigarette smoke exposure, recurrent infections, chronic pollution, or (less commonly) bad genetics, the lungs are constantly in an immune system battle with the environment in an effort to keep the system filled with clean air. Over time, as this battle continues to be waged, some of proteins that help maintain the shape of the alveoli (collagen and elastin) and the capillaries that feed the alveoli break down. Eventually, the alveoli fuse together, becoming larger, more cumbersome versions of their previous selves. Instead of 10 little balloons you end up with three or four much larger balloons. These balloons don't work as well and have difficulty getting air out. This decreases the surface area that's available for exchange, making diffusion harder! As more and more alveoli suffer from this change, the more and more difficult it becomes for the lungs to meet the body's demands for oxygen consumption and removal of carbon dioxide.

FUNCTIONAL HISTOLOGY OF THE CAPILLARIES AND THE ALVEOLAR−CAPILLARY UNIT

The alveolar membrane doesn't work by itself. When viewed as a unit that works in concert with the capillaries, the exchange membrane is actually made up of the alveolar cell wall, a very thick sliver of extracellular fluid, and then the capillary wall, as seen in Figure 2.4A. This figure demonstrates the extremely intimate contact between the capillaries and the alveolus. Considering the importance that both the alveolus and the capillary maintain their own independent integrity— the capillary needs to hold on to the blood that passes through, and the alveolus needs to hold on to the air that doesn't end up diffusing into the blood—the distance is about as small as you could ask for. Especially when you look at the figure and see that the capillary is so small that the red blood cell literally abuts right up against the wall, it becomes apparent that the distance that the oxygen and carbon dioxide molecules need to travel between red blood cell and alveolus is very minimal. If you take a look at Figure 2.4B however, you'll notice that even though the distance between the alveolus and pulmonary capillary is small, there are a number of elements that sit right in the middle. (There are actually 7 different layers between these two structures!)

Another interesting point about the histology of the capillaries that supply the alveoli is their density. They capillary beds are so dense that they have been described collectively as a "film" of blood continuously moving across the alveoli. This intimate alveolar−capillary relationship has led to the term alveolar−capillary unit to describe the two together,

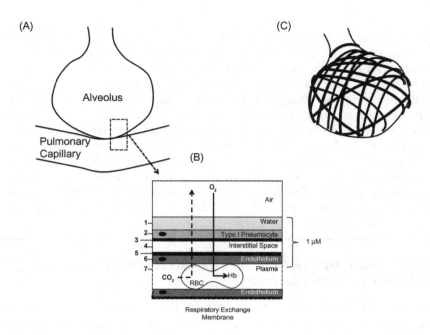

Figure 2.4 (A) Schematic representation of the alveolar capillary unit. Note the proximity of the alveolus to the capillary wall. (B) A close-up look at the respiratory exchange membrane between the alveolus and pulmonary capillary, which is approximately 1 μM thick and is composed of (1) Water lining the alveolar wall, (2) Type 1 Pneumocyte, (3) Pneumocyte basement membrane, (4) Interstitial Space, (5) Endothelial Basement membrane, (6) Capillary Endothelium, and (7) Plasma. (RBC − Red Blood Cell). (C) In reality, the density of capillaries around the alveoli is much greater, hence the name "alveolar−capillary unit." The flow of blood is so dense it's almost like a "blood film" covering the alveolar surface.

since they're histologically and physiologically linked. If you look at Figure 2.4C, you'll get a general feel for what this alveolar−capillary unit looks like in a slightly more three-dimensional aspect.

Because we wanted the focus of this book to be more on the oxygen and carbon dioxide and its movement and utilization into and within the body, we wanted to avoid a complex discussion about the cardiovascular system, both in terms of mechanics and anatomy. For more details on fluid mechanics of the body, please check out our first book on fluid physiology, *Back to Basics: Fluids in the Renal and Cardiovascular Systems*. It is important to note, however, that a high-pressure system would be bad if you're trying to incorporate new oxygen molecules from outside the environment via simple diffusion, right? If the blood pressure within the pulmonary capillaries were sky high, then there would be a tendency for fluid within the capillaries to leak out into the sliver of interstitial fluid noted in Figure 2.4A−B. It would then grow in size, and the increase in distance would have a negative impact on diffusion of oxygen and carbon dioxide. As it turns out, one of the particularly salient features of the pulmonary circulation—as opposed to the circulation of the rest of the body—is that it operates under extraordinarily low pressures and under very little resistance. In fact, the pressure is *literally* just enough to meet demand. If an adult is standing upright, there is *just* enough pressure to get blood flow uphill against gravity into the top parts of the lung. When called upon, however, it can accommodate a significant increase in blood flow in times of increased oxygen consumption such as during strenuous exercise. Note that we will talk at much greater length about the relationship between ventilation and blood flow in Chapter 6.

Clinical Correlate

Pulmonary Edema

Edema is Greek for "swelling." In medicine, it is known to be caused by an accumulation of fluid within a body compartment. In the case of pulmonary edema, it is the swelling of this interstitial space diagramed in Figure 2.4A–B. It can be caused by a host of things, including but not limited to abnormalities in the capillaries, viruses such as hantavirus, inflammation, strangulation, and sudden changes in altitude. Most commonly, it is caused by heart failure, specifically left-sided heart failure. Because the right side of the heart is responsible for pumping blood in the lower-pressure blood vessels feeding the lungs, and the left side of the heart is responsible for receiving the newly oxygenated blood returning from the lungs and pumping into the higher-pressure blood vessels feeding the rest of the body, a failure of the left side of the heart can cause a buildup of blood in the lungs. Because the low-pressure blood vessels of the lungs are not used to handling such a large amount of fluid (blood is predominantly made up of fluid), this fluid begins to leak out into the interstitial space via a pressure gradient. As this interstitial grows larger in size and the distance between the alveolus and the blood vessel becomes wider, it acts as an obstacle to the inward diffusion of oxygen. As such, patients with pulmonary edema can have significant problems oxygenating their blood.

CLINICAL VIGNETTES

Scenario 1

A 40-year-old man arrives in the Emergency Department after being shot multiple times in the right chest. He sustains massive trauma to the right lung and is taken to surgery where the decision is made to remove his right lung. He somehow makes it through and is discharged from the hospital 8 weeks later. He shows up about 6 months later for a follow-up clinic visit. He says he's been doing fine, but that he's short of breath after brief periods of light exercise.

1. How did removing this patient's lung affect diffusion?
   A. It changed the ΔP.
   B. It changed the surface area available for exchange.
   C. It changed the distance through which diffusion takes place.

Answer: B. By removing the right lung, this patient had an effective decrease of 50% of the surface area available for exchange. Neither the ΔP nor the distance for diffusion changed.

Scenario 2
A 67-year-old man with a history of heart failure presents to your office complaining that he is finding it increasingly difficult to breathe while lying down. He has a history of poor medication compliance. He also has been coughing more, and of late has been coughing up frothy, pink secretions.

1. The most likely reason for this patient's shortness of breath is most likely due to:
   A. A medication side effect from his sildenafil.
   B. Decrease in oxygen diffusion due to pulmonary edema.
   C. Pneumonia causing a V/Q mismatch.
   D. A change in partial pressure of oxygen due to inadequate ventilation.

Answer: B. The most likely answer here is B, pulmonary edema. Pulmonary edema due to left heart failure can, especially when more severe, present with pink (blood-tinged), frothy secretions. The shortness of breath while lying down is known as *orthopnea*, and occurs simply because a larger percentage of the alveoli are affected by the edema as it follows gravity. If you're laying down flat on your back, rather than just sitting at the bases of your lung, the fluid will begin to occupy the entire back (posterior) portions of your lung, causing a larger number of alveolar–capillary units to be affected.

CHAPTER 3

Lung Mechanics: Putting the Blueprints of Gas Exchange into Action

As we've stated multiple times now, diffusion is the driver behind all movement of gases in the body. *Ultimately, what keeps oxygen moving into the cells and carbon dioxide out of the cells is diffusion!* And, all of this diffusion takes place collectively through the hundreds of millions of alveolar–capillary units. Remember that in the end, O_2 delivery and CO_2 removal is only as fast as its diffusion gradient allows! Diffusion, as you recall from the diffusion equation (which, by the way, is tested on STEP 1 USMLE Board Exams), is improved by (1) increasing concentrations of particles on one side of the membrane, that is, increasing the ΔP; (2) increasing membrane surface area; and (3) decreasing distance between compartments. Intrinsic to the blueprints themselves, to the anatomy, are (2) and (3). The distance and surface area are intrinsic to the system. Delta P, however, is what happens once the movement— the mechanics of the lungs—gets turned on. All of these variables play a role in the continuous exchange of O_2 and CO_2, but the ΔP is the driving force behind gas exchange. If there is no pressure difference, it does not matter how small the distance or how big the surface area available. This is why the renewal of alveolar air with fresh atmospheric air is critical! Keeping the alveolar partial pressure of O_2 (P_AO_2) higher, and the alveolar pressure of CO_2 (P_ACO_2) lower than the partial pressures inside the body keeps the ΔP stable. By keeping ΔP stable, you keep oxygen flowing in and CO_2 flowing out.

Key

Alveolar air must be continually renewed with atmospheric air in order to allow for diffusion of both O_2 and CO_2.

Given that the ΔP is so important, let's take a closer look at the actual numbers involved in the movement of O_2 and CO_2. Remember from Chapter 1 that atmospheric pressure at sea level is 760 mmHg.

Back to Basics in Physiology. DOI: http://dx.doi.org/10.1016/B978-0-12-801768-5.00003-4

This number is a result of the added partial pressures of all the gases that make up the atmosphere. Thus, atmospheric air is made up of approximately 78% nitrogen (N_2), 21% oxygen (O_2), and trace amounts of other gases including carbon dioxide (CO_2).

Take a look at Table 3.1 and you'll be able to see how these percentages translate into actual pressures. On an average day, in the atmospheric air, the pressure of O_2 is approximately 160 mmHg. As you breathe the air in, however, there's a sharp decrease in the pressure of O_2 to 150 mmHg. This is because normal body temperature of 37°C generates water vapor. This water vapor has a partial pressure of 47 mmHg and will dilute the rest of the gases in that fraction of air, therefore PO_2 will decrease to 150 mmHg and PN_2 will decrease to 563 mmHg. As air finally reaches the alveoli, there's an even bigger drop in alveolar PO_2 (P_AO_2) to 104 mmHg because oxygen is being consumed. The PCO_2 in the alveoli (P_ACO_2) is approximately 40 mmHg, and as we will see later, this is closely regulated. An alveolar PO_2 of 104 mmHg and P_ACO_2 of 40 allows for an adequate exchange of O_2 and CO_2 with blood, and ensures oxygen continues to move in and carbon dioxide out. Remember that these gases continue to be used in every cell of the body as the cells are constantly consuming oxygen and producing carbon dioxide. Thus if the alveoli are not constantly receiving fresh atmospheric air, this gradient, this ΔP, will be wiped out! Eventually, most of the O_2 would be sucked out of the alveoli, consumed by the cells and replaced by CO_2. Therefore, renewal with fresh air is essential to sustaining life.

In this chapter and the next, we will see what happens once the blueprints are put into action, the mechanics of ventilation. This chapter will be focused on understanding the methods employed to generate

Table 3.1 Approximate Standard Partial Pressures of Gases at Sea Level on an Average Day				
Partial Pressures of Gases in mmHg				
	Atmospheric Air	**Moist Tracheal Air**	**Alveolar Air**	**Expired Air**
PO_2	160	150	104	120
PCO_2	0	0	40	27
PH_2O	0	47	47	47
PN_2	600	563	569	566
Ptotal	760	760	760	760

air movement within the lungs and the physics behind it. The next chapter will focus on what this looks like when integrated more fully with the human anatomy—the respiratory cycle. So let's get started!

IN AND OUT: HOW GAS MOVES

The goal of the lungs is to "mix" the gases coming from the true atmosphere and the gases coming from the cells in the body. This aids in maintaining the diffusion gradients for O_2 and CO_2. But, believe it or not, the lungs are relatively passive bystanders in all of this. How so? Well, the lungs themselves cannot actively contract or expand. Some of the conduction airways have smooth muscle in them, but as a whole the lungs don't. If the lungs cannot actively contract or expand themselves to move air in and out, it means that it is mainly external forces—that is, movement of the diaphragm, the chest wall, surface tension, and so on—that are acting upon the lungs to either expand them or contract them. This is an extremely important concept that is bypassed regularly by many students as obvious; however, the implications are far reaching in both health and disease. In order to understand this a little better, let's first study the relevant anatomy from a functional and mechanical point of view.

Key
The expansion and contraction of the lungs is passive and it depends on external factors such as the diaphragm, chest wall motion, surface tension, and lung tissue elastic recoil.

Functional and Mechanical Anatomy

A simple analogy for lung function would be that the lungs are one big balloon (actually many hundreds of millions of interconnected little balloons, but the same logic can be applied). So, how does inflating a balloon work? Well, let's take a look. Figure 3.1 shows us the basics of the pressure differentials required to inflate a balloon. The principle to inflating the balloon is:

Establish a pressure gradient so that there is a pressure force driving air from the mouth of the balloon (high pressure) to the inside of the balloon (low pressure). This will begin to increase the pressure within the balloon until the gradient is lost and the pressure on both sides is equal.

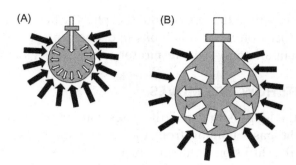

Figure 3.1 When the inflation pressure (white arrows) is less than the sum of the elastic resistance of the balloon and the pressure exerted by the atmosphere (black arrows) the balloon will not inflate (A). However, when the inflation pressure overcomes both elastic resistance and atmospheric pressure, the balloon inflates (B).

In addition to establishing a pressure gradient from the mouth of the balloon to the inside of the balloon, two things need to be overcome first in order to drive air into the balloon:

- The elastic resistance of the balloon itself
- The pressure exerted on the balloon by the atmosphere

In Figure 3.1A, the black arrows represent the pressure working against the balloon being inflated (in this case elasticity and outside atmospheric pressure). The white arrows represent the pressure inside that is trying to inflate the balloon. Remember, pressure is just force over a given area, so what we're looking at is a balance of forces. It is the imbalance between these forces that will determine whether or not the balloon inflates. This means that when the inflation pressure is greater than both the elastic resistance and the atmospheric pressure, the balloon will inflate (see Figure 3.1B). If the inflation pressure is not large enough to overcome these opposing forces, then the balloon won't inflate. Simple enough, right?

The Importance of the ΔP: It's Not Just at the Alveoli, but Getting to the Alveoli as Well

What will ultimately move air in and out of the lungs is a pressure difference between the lungs and the atmosphere. So, let's analyze this phenomenon with a two-compartment model like that shown in Figure 3.2A. In it, we can see two compartments communicating via a channel between them. These compartments are at sea level, so the pressure of air inside them is 760 mmHg. Now, the channel that communicates both compartments is open, thereby the pressure of gas in both compartments equalizes

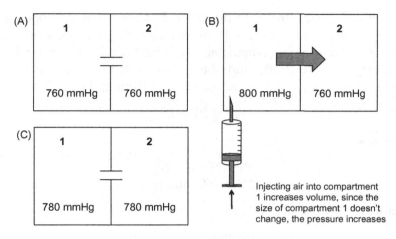

Figure 3.2 The movement of gas from one compartment to another requires the presence of a pressure gradient. At a steady state, where the pressure is the same in both compartments (A), no air will move. By increasing the pressure in compartment 1 (B) a gradient of +40 mmHg is generated from compartment 1 to 2, which favors diffusion of gas until the pressure is again equalized between compartments (C).

and is therefore the same; that is, there is *no* pressure gradient. Put as a formula, in order to calculate the ΔP between compartments:

$$\Delta P = P1 - P2$$

in this case:

$$P1 = 760 \text{ mmHg}$$
$$P2 = 760 \text{ mmHg}$$

therefore:

$$\Delta P = 760 \text{ mmHg} - 760 \text{ mmHg}$$

So:

$$\Delta P = 0 \text{ mmHg}$$

Key

Δ (*delta*) = difference, so ΔP means the difference in pressure between two compartments and ΔVol refers to the difference in volume between two compartments.

A ΔP of 0 mmHg means that there is no pressure difference between compartments, so there is no net gradient for movement of gas from one compartment to the other. But what would happen if we injected air into compartment 1 without changing the size of the container? The pressure in compartment 1 would rise from 760 mmHg to 800 mmHg (Figure 3.2B) and the pressures between compartments would no longer be equal. If we use our ΔP formula again:

$$\Delta P = P1 - P2$$

in this case:

$$P1 = 800 \text{ mmHg}$$
$$P2 = 760 \text{ mmHg}$$

therefore:

$$\Delta P = 800 \text{ mmHg} - 760 \text{ mmHg}$$

So:

$$\Delta P = +40 \text{ mmHg}$$

The ΔP is now positive! Which means that air will flow from compartment 1 into compartment 2 until the pressures are equalized. In this case, let's say that the pressures equalize at 780 mmHg (Figure 3.2C). If we now plug these numbers into our formula, you'll see that the ΔP is now 0 again. This is because even though total pressure of our system increased (780 mmHg vs 760 mmHg), there is no net pressure difference between compartments.

Now, back to the balloon. This time, we placed the balloon inside of a bottle (Figure 3.3A). Note that the mouth of the balloon is open and attached to the opening of the bottle. So, similar to what we saw in Figure 3.2, we have two compartments in our balloon-in-a-bottle model:

- The compartment outside of the balloon (the atmosphere), Compartment 1
- The compartment inside the balloon, Compartment 2

It's important to note that the inside of the bottle is *not* communicating with either the outside atmosphere or the balloon (greyed out area in Figure 3.3). It is sealed off. In Figure 3.2 we had two freely communicating compartments, so the pressures between them always equalized. In Figure 3.3, the atmosphere and the balloon are freely

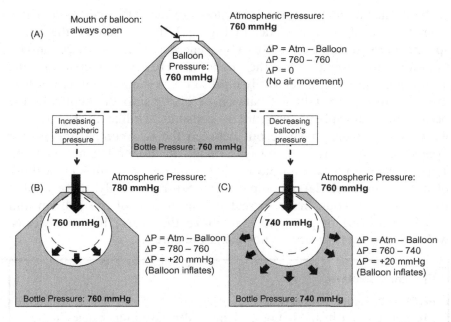

Figure 3.3 A simplified model of lung function. At equilibrium (A), there is no net movement of air. The pressure inside the balloon (white) will always attempt to reach equilibrium with the atmospheric pressure. Increases in atmospheric pressure (B) or decreases in bottle pressure (C) will both generate a gradient that favors balloon inflation.

communicating because the mouth of the balloon is open. So, if the atmospheric pressure in Figure 3.3A is 760 mmHg, what do you think the pressure will be inside the balloon? Exactly the same: 760 mmHg! (Remember, the pressures in compartments that are freely communicated will tend to equalize.) This means that the ΔP between the atmosphere (760 mmHg) and the balloon (760 mmHg) is 0. Therefore, there will be no net movement of air in or out of the balloon, because there is no gradient between the atmosphere and the inside of the balloon.

Our ultimate goal is to inflate the balloon. In order to do that the pressure inside the balloon needs to be less than the pressure outside. Therefore if we plug this into our formula for ΔP, then:

$$\Delta P = \text{Atmospheric Pressure} - \text{Balloon Pressure}$$

If we want the ΔP to be positive (and have the balloon inflate) we can do one of two things:

- We can increase the atmospheric pressure.
- We can decrease the pressure inside the balloon.

If we increased the atmospheric pressure the ΔP would be positive from the outside to the inside. In Figure 3.3B we increase the atmospheric pressure to 780 mmHg. This generates a ΔP of +20 mmHg, favoring air going into the balloon. However increasing the atmospheric pressure is relatively hard to do. Assuming you're not hiking up a mountain with the balloon or diving into the depths of the ocean, the atmospheric pressure is constant. Unlike in Figure 3.2, where the pressures average out between the two compartments, the atmospheric pressure will always remain at 760 mmHg at sea level. This is because the earth atmosphere is so large, you can almost think of it as an infinitely large compartment. So, realistically, if we wanted to do this, we would have to seal off the mouth of the balloon and attach some sort of pump to it to make this work. In other words, we'd have to create a new atmospheric pressure.

Clinical Correlate

Positive Pressure Ventilation

Movement of air in and out of the lungs is dependent on changes in pressure that favor a decreased pressure inside of the lungs relative to the atmosphere to bring air in, and an increased pressure inside the lungs relative to the atmosphere to let air out. However when a patient can't breathe on his or her own, we use something called Positive Pressure Ventilation (PPV). PPV does just what its name describes. Generally this is done through an endotracheal tube (tube that is placed in the trachea) that is connected to a ventilator, which moves air in and out of the lungs. To get air into the lungs it generates a positive pressure in the ventilator (so the pressure inside the lungs is less), which generates a gradient from the outside in. To get air out of the lungs, the reverse happens. The pressure in the ventilator is decreased allowing air to flow out of the lungs.

But we want to try and get this balloon to inflate without some sort of external pump. So what other option do we have? Well, we can try to decrease the pressure inside the balloon relative to the atmosphere. Decreasing the pressure inside the balloon would still create a ΔP. But how would you do this? This is where the bottle that we've so casually ignored up to this point comes in.

Given that the balloon is inside the bottle, and the bottle is effectively sealed from the outside, the pressure inside the balloon will be a direct reflection of the pressure inside the bottle. Consequently if we

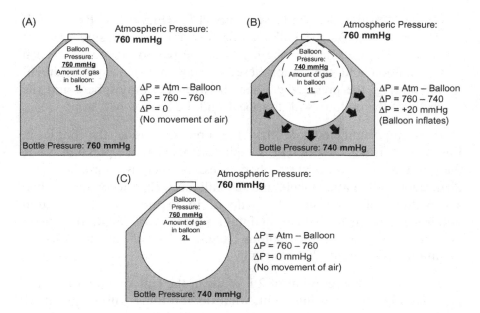

Figure 3.4 The decrease in bottle pressure (A to B) will initially lead to a transient decrease in balloon pressure (B). The decreased balloon pressure will generate a pressure gradient from the outside in. Air will flow down the pressure gradient and fill the balloon with air until the atmospheric pressure and balloon pressures are equal again (C).

decrease the pressure inside the bottle, this creates a gradient relative to the outside, which will in turn decrease the pressure inside the balloon (Figure 3.3C). Think about it like this: In Figure 3.3B the positive atmospheric pressure pushes the balloon open and fills it with air (the black arrows are inside the balloon pushing out). In Figure 3.3C, the pressure inside the bottle is less than atmospheric pressure. That makes it *by definition* a vacuum! It therefore acts as a vacuum in "pulling" the balloon open from the outside (the arrows are outside the balloon pulling its walls open). By pulling on the balloon's walls, two things happen in sequential order (Figure 3.4):

1. The size of the balloon increases (i.e., the balloon gets bigger). This will decrease the air pressure inside the balloon because the same amount of air is distributed in a larger space, therefore in order to equalize the pressure with the outside you need air to flow in (Figure 3.4A, B).
2. Once the balloon starts to get bigger and the pressure inside begins to decrease, air starts flowing into the balloon in order to equalize the pressures. The air flow into the balloon will increase the volume

of gas inside the balloon from 1 L (Figure 3.4B) to 2 L (Figure 3.4C). As the balloon fills with air, the pressure gradient between the outside and the inside of the balloon decreases. When the balloon and the atmosphere once again reach equilibrium at 760 mmHg, the pressure gradient between the atmosphere and the inside of the balloon will disappear and the airflow will stop.

It is important to note that the pressure inside the bottle in Figure 3.4C is still 740 mmHg. In spite of this negative pressure inside the bottle relative to the atmospheric pressure, there is no pressure gradient between the atmosphere and the balloon. What accounts for this? The volume of gas inside the balloon! In Figure 3.4A the balloon requires 1 L of air in order to maintain a pressure of 760 mmHg, however in Figure 3.4C, the balloon requires 2 L of air in order to maintain a pressure of 760 mmHg.

Taking this new set point of 2 L of air in the balloon and a pressure of 760 mmHg as a starting point, what do you think will happen if we further decrease the pressure inside the bottle? Let's take a look. In Figure 3.5A all we've done is taken our balloon with 2L and pasted it there, nothing has changed. The pressure inside the balloon and the atmosphere is the same (760 mmHg). Therefore there is no pressure gradient and air will not be moving. If we decrease the pressure in the bottle to 735 mmHg (Figure 3.5B), what do you think is going to happen? The same thing that happened before when we decreased the pressure in the bottle—the pressure in the balloon will decrease accordingly, generating a pressure gradient from the outside in. This will make the balloon inflate in order to equalize the pressures between the atmosphere and the balloon once again. In this case the volume of air inside the balloon rose by 500 mL to 2.5 L. Once the volume of air in the balloon increases to 2.5 L the pressures equalize and there no longer is movement of air.

What will happen if we then increase the bottle pressure from 735 mmHg back to 740 mmHg? The exact opposite—the pressure inside the balloon will increase to higher than the atmospheric pressure and air would leak out of the balloon until equilibrium was once again reached at 760 mmHg, which would mean moving back to Figure 3.5A. If we were to graphically plot out the changes in volume of air inside the balloon we would get something like Figure 3.5C, where you can see time (sec) on the X-axis and volume of air on the Y-axis. As air goes into the balloon there would be an upstroke (white arrow) in the volume, and as

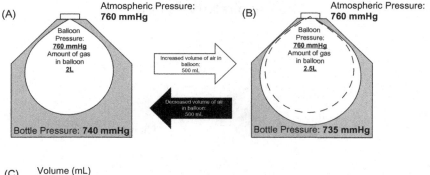

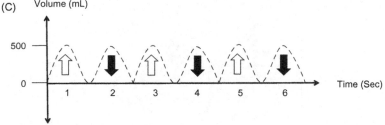

Figure 3.5 Changes in bottle pressure from A to B and then from B to A will lead to a cyclical movement of air in and out of the balloon (C).

air then leaves the balloon, a downstroke (black arrow). So by doing this several times in a row, what we're effectively doing is moving air in and out of the balloon by changing the pressure in the bottle. I think you know where we're going with this: The lungs work in a way that is very similar to our balloon in a bottle, where the bottle represents the chest cavity and the balloon represents the lungs.

PLEURAL PRESSURES: NEGATIVE VERSUS POSITIVE PRESSURE

The lungs and the chest work in a way that is very similar to the balloon-in-a-bottle model that we've been discussing thus far. If you look at Figure 3.6, you'll see some basic anatomy of the respiratory system. In our model, the mouth of the balloon/bottle is the glottis and trachea. The balloon represents the lungs. The bottle is the thoracic cavity. If you imagined the entire inside surface of the bottle and the outside surface of the bottle covered in one contiguous film or membrane, it would be a good approximation of the pleura. The lungs and the pleura hang inside the thoracic cavity and communicate with the outside via the trachea. The glottis acts as the door to the trachea, and can open

The Respiratory System

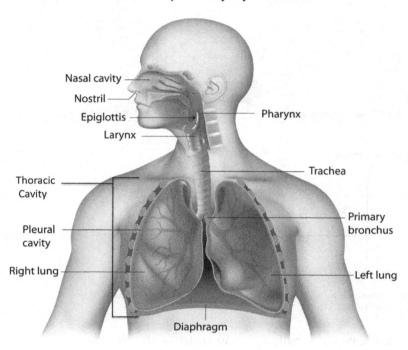

Figure 3.6 Anatomical representation of the ventilation system.

and close. Similarly, the remaining upper airway structures also have some ability to open and close (your mouth, e.g.). If you take a look at Figure 3.7, you'll see a simplified diagram of the relevant anatomy. In it you can see that the pressure inside the chest is negative!! This negative pressure is the one responsible for "propping open" the lungs with air from the outside. How is the negative pressure achieved within the chest? Again referring to Figure 3.7, you'll see that the lungs (white) are hanging inside the pleural space (greyed out area). The pleural space is defined as the space between two very thin layers of cells that cover the outside of the lungs (visceral pleura) and that line the inside of the chest wall (parietal pleura). In order to generate negative pressure in the pleural space, the lymphatic system is constantly removing fluid from the pleural space. Through this constant removal of fluid, the pressure inside the pleural space is negative compared to the outside. In general, the pleural pressure is said to be a pressure of -3 to $-5\,\mathrm{cmH_2O}$. Does this mean that the pressure inside the pleural space is a true vacuum?

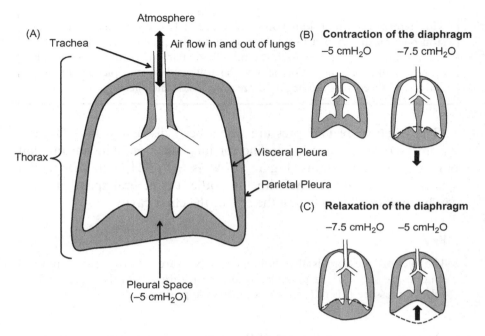

Figure 3.7 Simplified anatomical representation of the lungs and the thoracic cavity (A). When the diaphragm contracts (B) it gets shorter and thereby increases the size of the pleural space. The increased size of the pleural space decreases the pressure from −5 cmH₂O to −7.5 cmH₂O. Conversely, when the diaphragm relaxes it gets longer and decreases the size of the pleural space, thereby increasing the pleural pressure from −7.5 cmH₂O to −5 cmH₂O (C).

Not at all, what this means is that relative to atmospheric pressure the pleural space has a pressure that is slightly less. This "less than atmospheric" pressure is enough to bring air into the lungs!

Clinical Correlate

Pneumothorax

In order to keep the lungs inflated the pleural space needs to have negative pressure. There are certain clinical situations in which there can be a hole in either the parietal pleura (e.g., a rib fracture or stab wound) or in the visceral pleura (e.g., tear in the lung from a burst bullae). These holes allow for air to leak into the pleural space, which collapses the lung. The clinical name for this is a "pneumothorax" or "air in the thorax." If the air leak is caused by a wound that acts as a sort of one-way valve that lets air in but doesn't let it out, massive amounts of air can build up inside of the chest cavity. This type of pneumothorax is called a "tension pneumothorax," and it is a medical emergency because it can compress the heart and great vessels in the chest and cause circulatory collapse.

This is precisely why it kills you if left untreated. The treatment for a tension pneumothorax is a rapid decompression by puncturing a hole in the second intercostal space along the mid-clavicular line of the affected side to relieve the pressure. This should be followed by the placement of a chest tube to correct the negative pressure inside the chest.

The negative pleural pressure we've been discussing thus far (i.e., $-5\,cmH_2O$) serves as our baseline for lung function. This means that our starting point (think Figure 3.7A) is $-5\,cmH_2O$ in the pleural space. And just as we saw with the bottle, the pleural space pressure can be modified by changing the size of the chest cavity.

Key

The pleural pressure is the difference in pressure relative to the atmospheric pressure. If the pressure inside the pleura is less than the atmospheric pressure it is referred to as a negative pressure.

In order to change the size of the chest cavity, we must use the muscles of respiration. These are specialized muscles, which change the size of the thorax as they relax and contract. The most important muscle of ventilation is the diaphragm, which lies just below the lungs and divides the thorax from the abdomen. As the diaphragm contracts, it gets shorter. As it gets shorter it moves down into the abdomen and makes the chest cavity larger! If we increase the size of the chest cavity as in Figure 3.7B, the pressure in the pleural space will decrease from $-5\,cmH_2O$ to $-7.5\,cmH_2O$. Therefore, contracting the diaphragm will decrease the pressure inside the lungs, which makes air flow in (see how the lungs get bigger in Figure 3.7B). Conversely in order to increase the pressure within the chest and force air out, the diaphragm needs to relax (get longer), and as it does this it rises back up, effectively shrinking chest cavity. This will increase the pressure in the chest cavity from $-7.5\,cmH_2O$ to $-5\,cmH_2O$ and air will escape the lungs into the atmosphere (Figure 3.7C). These are the basic tenets of how air moves in and out of the lungs using negative pressure. However, before we completely address the specific pressures of the lung and the chest we need to address two very important concepts: compliance and elastance, since they will be key determinants of the amount of air that moves in and out of the lungs.

LUNGS OUTSIDE THE BODY: TISSUE DYNAMICS

As we've briefly reviewed, the lungs expand and contract in response to changes in the size of the thoracic cavity. But how the lungs respond to changes in the pleural pressure depends on the tissue characteristics of the lungs themselves. So, let's take a quick look at some of the properties of the lung tissue that define how the lungs expand and contract.

Compliance and Elastance

When understanding lung mechanics, there are two concepts that are absolutely key: compliance and elastance. Compliance is defined as change in volume divided by change in pressure, or in other words: How much does the volume change in response to a change in pressure? In equation form:

$$\text{Compliance} = \frac{\Delta\text{Volume}}{\Delta\text{Pressure}}$$

Keeping with our previous example, if a balloon is highly compliant (Figure 3.8A), a small change in pressure will generate a very large change in volume. If the balloon is not compliant (Figure 3.8B), the same change in pressure as before will generate only a small change in volume. If we were to plot this as a graph, the compliance is the slope of the relationship between volume and pressure. (Remember from algebra that you get the slope by plotting $\frac{\text{rise}}{\text{run}}$; in this case $\frac{\text{rise}}{\text{run}} = \frac{\Delta\text{Volume}}{\Delta\text{Pressure}}$). This means that the slope represents the compliance. So, if we were to calculate the compliance of **A** and **B**, we would get the following.

Compliance of balloon in Figure 3.8A

$$= \frac{6L - 1L}{0\,\text{cmH}_2\text{O} - (-5\,\text{cmH}_2\text{O})} = \frac{5\,L}{5\,\text{cmH}_2\text{O}} = 1\,\text{L/cmH}_2\text{O}$$

Compliance of balloon in Figure 3.8B

$$= \frac{2L - 1L}{0\,\text{cmH}_2\text{O} - (-5\,\text{cmH}_2\text{O})} = \frac{1L}{5\,\text{cmH}_2\text{O}} = 0.2\,\text{L/cmH}_2\text{O}$$

Take a look at Figure 3.8C and you'll be able to see just this. The graph depicts the compliance of both balloons plotted against each other. It very clearly shows that with the same ΔPressure (5 cmH$_2$O) the ΔVolume is higher for A (more compliant) and lower for B (less compliant).

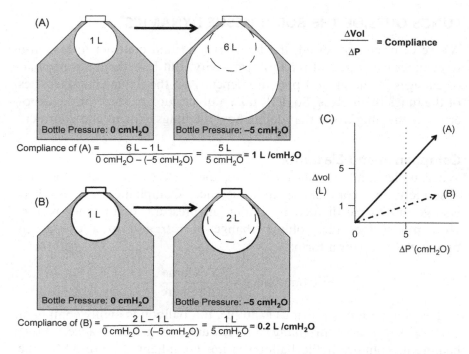

Figure 3.8 Compliance is determined as the ΔvolΔpressure. The balloon in (A) is more compliant than that of (B) because with the same Δpressure, the volume in (A) increases 5 times as much as the volume in (B). A graphical representation of the compliance of both balloons (C), clearly shows how A is more compliant than B.

Key

Compliance is defined as ΔVolume/ΔPressure, and it defines how much the volume will change in response to changes in pressure.

In Figure 3.8, the reason behind the difference in compliance between the two balloons is very easy to identify! (Just in case you haven't, look at the thickness of the balloon wall.) The balloon in Figure 3.8A has a wall that is very thin relative to that of the balloon in Figure 3.8B. Therefore we can say that something that is more compliant is more easily stretched out, while something that is less compliant is harder to stretch out. Along the same lines, something that is easily stretched out is less likely to recover its original shape, whereas something that opposes stretch is more likely to recover its original shape. The property of opposing stretch is called elastance and it's the reciprocal of compliance.

$$\text{Elastance} = \frac{\Delta\text{Pressure}}{\Delta\text{Volume}}$$

This means that the more compliant something is, the less elastic it will be, and the more elastic something is, the less compliant it will be. So, going back to Figure 3.8, the balloon in **A** is very compliant and not very elastic, whereas the balloon in **B** is not very compliant and very elastic. Why is this important? Understanding compliance and elastance is key in understanding lung mechanics because the compliance and elastance of the lungs determine how much they will inflate or deflate with changes in pleural pressure. The relationship between compliance and pleural pressure is the focus of our next subsection.

Transpulmonary Pressure

Thus far, whenever we've changed the pressure inside the bottle (e.g., Figures 3.4, 3.8) the change of pressure inside the balloon has been identical (i.e., if we decreased the pressure inside the bottle by 5 cmH$_2$O the pressure in the balloon changed by the same 5 cmH$_2$O). This would mean that absolutely no pressure is lost on the rubber of the balloon itself. This is not entirely accurate. In order to inflate the balloon, some of the pressure will be lost on the rubber that makes up the balloon; that is, the rubber that composes the balloon is also going to have a resistance to stretching that needs to be overcome in order to inflate the balloon. In Figure 3.8 we saw two balloons that had different compliances. We mentioned only that it had to do with the wall thickness. If you take another look at these balloons, though, you'll see the increased wall thickness of balloon B makes it less compliant. Since balloon B is less compliant, it means that for the same change in pressure (in this case, -5 cmH$_2$O) the change in volume is small compared to a more compliant balloon A. If we wanted to increase the volume in balloon B to a full 5 L we would theoretically require a pressure of -25 cmH$_2$O. That's five times as much pressure for balloon B as for balloon A— what the heck is all the extra negative pressure doing? The simple answer is that it's holding the wall of the balloon open, so the thicker the wall, the more pressure that's going to be required.

The lungs work in a very similar fashion. As we previously discussed, the lungs are made up of multiple types and layers of tissue with the sole purpose of getting air to the alveolar sac where it can be used in the diffusion of O$_2$ and CO$_2$. All these tissue fibers, and predominantly collagen and elastin, are going to behave similarly to the wall of the balloon we've been discussing. Figure 3.9A shows us the lungs sitting inside the pleural space, but inside the lungs you will see a gross representation of the alveolar sacs. Since it's in the alveolar sacs where exchange takes place, it's in the alveolar sacs where the pressure needs to decrease to allow air to flow in and increase to push the air out. So keep in mind that what we're really interested in is the change in pressure at the level of the alveolar sacs.

Take a look at Figure 3.9B; it's the same schematic as Figure 3.9A, but this time, we've included the ΔP in cmH_2O relative to the atmosphere. So, this means that if the pressure is 0 cmH_2O (it doesn't mean that the pressure in the alveoli is a vacuum), it means that the ΔP between the alveoli and the atmosphere is 0 cmH_2O. In other words, it is the same pressure in the alveoli as in the atmosphere, so there is no net pressure gradient for the movement of air. If 0 cmH_2O is our baseline, when we mention that the pressure in the alveoli becomes negative that means that it is lower than the atmosphere and air will flow into the lungs. When the pressure in the alveoli becomes positive it means that the pressure in the alveoli is higher than the pressure in the atmosphere and therefore air will flow out of the lungs.

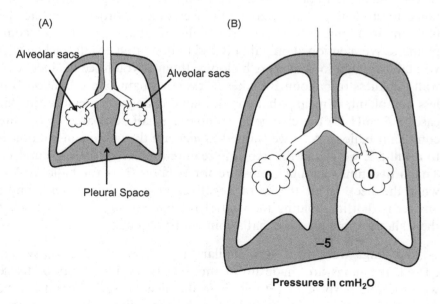

Figure 3.9 The alveolar pressure will always try to equilibrate with the atmospheric pressure. Once equilibrated there will be NO pressure gradient between the alveoli and the atmosphere; i.e., the Δ pressure will be 0 cmH_2O.

Let's take a look at Figure 3.10. This is a busy figure, so we'll walk you through it step by step. Across the top are schematics for different alveolar pressures and the corresponding pleural pressures. The sequence of events that we're about to analyze happens almost simultaneously. However, in order to make the progression from one step to the other a little clearer, we'll break down each step to its simplest elements. The lungs on the left (Figure 3.10A) are the starting point, with a baseline pleural pressure of -5 cmH$_2$O and an alveolar pressure difference of 0 cmH$_2$O. Remember a ΔP of 0 cmH$_2$O in the alveoli means that it's the same as atmospheric pressure. In order to bring air into the lungs, the pressure in the alveoli needs to decrease. This is achieved by increasing the size of the pleural space (like we studied in Figure 3.7). In Figure 3.10B you can see how by increasing the volume of the pleural space, the pleural pressure has gone from -5 cmH$_2$O to -7.5 cmH$_2$O. This sudden decrease in pleural pressure will decrease the alveolar pressure difference from 0 cmH$_2$O to -1 cmH$_2$O, which means that the alveolar pressure is now 1 cmH$_2$O *lower* than the atmospheric pressure. Given this pressure difference, air will begin to flow

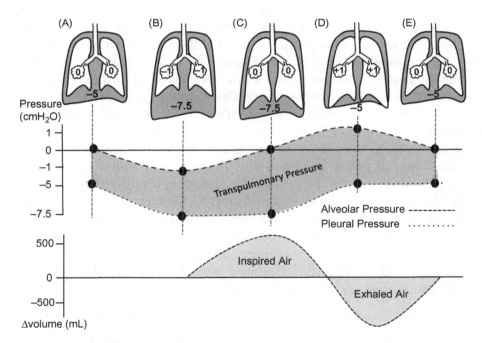

Figure 3.10 The respiratory cycle. Sequential changes in the pleural pressure lead to changes in the alveolar pressure. When the alveolar pressure is negative with respect to the atmospheric pressure, air flows into the lungs. When the alveolar pressure is positive with respect to the atmospheric pressure, air flows out of the lungs.

into the lungs and they will inflate (Figure 3.10C). As the lungs get bigger (Figure 3.10C), the pressure difference between the atmosphere and the alveoli disappears. Even though the pleural pressure is still $-7.5\,cmH_2O$, the alveolar pressure is now $0\,cmH_2O$ due to the increase in lung volume.

So that's how you get air in. But how do you push the air out? By reversing the alveolar-atmosphere pressure gradient. This is achieved by increasing the pleural pressure from $-7.5\,cmH_2O$ to $-5\,cmH_2O$ (Figure 3.10D). As the pressure inside the pleural space increases, the excess volume inside the lungs will be compressed, which will in turn increase the pressure in the alveoli to $+1\,cmH_2O$ relative to atmospheric pressure. This pressure difference now favors the exit of the air from the lungs into the atmosphere (Figure 3.10E) and a subsequent return to baseline volume and pressure.

All the concepts that we just discussed are presented in graphical format in the bottom-most part of Figure 3.10. Let us first draw your attention to the bottom graph. In it you can see the changes in volume (Y-axis) that are associated with changes in pressure (X-axis) within the lungs. And, as we just saw, when the alveolar pressure is lower than that of the atmosphere (Figure 3.10B), air will begin to flow in until the pressure is equalized (Figure 3.10C). In this particular case, the $-1\,cmH_2O$ pressure difference moves 500 mL of air into the lungs (shaded area labeled inspired air). Conversely, a $+1\,cmH_2O$ increase in alveolar pressure pushes 500 mL of air out of the lungs (shaded area labeled exhaled air). Notice that, in the balloons with which we had been working, the change in pressure in the balloon was identical to the change in pressure in the bottle. In the lungs however, the change in pleural pressure is *larger* than the change in alveolar pressure; that is, the pleural pressure changes $2.5\,cmH_2O$ for every $1\,cmH_2O$ change in alveolar pressure. This means that you need a larger pressure difference to expand or contract the lungs. This difference in pressure is called the *transpulmonary pressure*, and literally means the pressure across the lung. It is the measured pressure difference between the pleural pressure and the alveolar pressure.

Transpulmonary Pressure = Alveolar Pressure − Pleural Pressure

Let's analyze this a bit further. The alveolar pressure in both Figures 3.10A and 3.10C is the same (0 mmHg) but the transpulmonary pressure is different. Figure 3.10A has a transpulmonary pressure of -5 cm H_2O, and in Figure 3.10C it's -7.5 cm of H_2O. What's going on? Well, look at the volume of both lungs. The volume of Figure 3.10A is less than the volume of Figure 3.10C. Therefore, we can say that the pressure in the alveolus is equal to the pressure in the atmosphere; that is, the pressure gradient is 0 cmH_2O. The transpulmonary pressure is the amount of pressure we need to keep the lung volumes constant. If the transpulmonary pressure remains constant there will be no change in lung volume. It is the changes in transpulmonary pressure that modify alveolar pressures, but what exactly is transpulmonary pressure a function of?

Lung tissue naturally opposes stretching (this is a function of both tissue resistance and surface tension), and transpulmonary pressure is a measure of how much resistance the lung poses to stretching. This concept is directly related to compliance! The more compliant a lung is, the less pressure you need to change the volume. Therefore a lung that is highly compliant will require a SMALL transpulmonary pressure to inflate; it has a small resistance to stretching, which means you only need a small pressure gradient to stretch it. Meanwhile, a lung that is not compliant, which requires a lot of pressure to change the volume, will require a very LARGE transpulmonary pressure to inflate. How is this even remotely relevant to clinical medicine, you ask? Well, if the lung is very stiff and requires a large transpulmonary pressure, it's going to be difficult to change the pressure in the alveoli and therefore extremely difficult to inflate the lungs! So diseases that alter lung compliance will affect transpulmonary pressure, which will have a direct effect on lung function.

How exactly can disease alter lung compliance? There are two particular disease states that can modify lung compliance: emphysema and fibrosis. Emphysema destroys the collagen and elastin fibers in the lung, thereby making the lungs highly compliant (floppy), whereas fibrosis deposits fibrous tissue that stiffens the lung and makes them less compliant (stiff). Figure 3.11 is a graph depicting the differences between normal lung compliance (**A**), the increased compliance of an emphysematous lung (**B**), and the decreased compliance of a fibrotic lung (**C**). The differences in compliance are clearly evident because the same change in pressure (dotted line) has very different effects on the change in volume in each of the lungs. The more compliant emphysematous

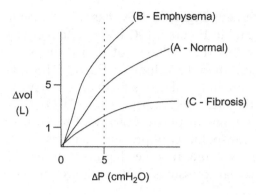

Figure 3.11 Lung compliance curves for (A) normal lungs; (B) Emphysematous lungs, which have an increased compliance (more Δvol); and (C) Fibrotic lungs, which have a decreased compliance (less Δvol) to the same pressure.

lung will have a much larger change in volume than the normal lung or the fibrotic lung, which will only have a modest increase in volume. This is to say, the more compliant a lung is, the easier it is to inflate, whereas the less compliant it is, the harder it is to inflate.

Clinical Correlate

Fibrosis and Emphysema

Lung compliance is a key factor that determines how lungs will behave under different changes in pressure. Emphysema, which destroys the collagen and elastin fibers in the lung, makes the lungs highly compliant, whereas fibrosis, which stiffens the lungs, makes them less compliant. As you can imagine, the clinical scenarios between both conditions are different. One lung has trouble getting air in (fibrosis), while another has trouble getting air out (emphysema)!

Aside from intrinsic tissue resistance, there are two other very important factors that play a role in lung mechanics: alveolar radius and surface tension. However, in order to visualize how radius and surface tension play a role in lung mechanics, we need to graph a compliance curve for the entire ventilation system (i.e., lung and chest wall put together). But to do that we need to understand how the compliance of the entire system is calculated. In Figure 3.11, we saw what the compliance of the lungs outside of the chest cavity looks like. Let's begin by plotting normal lung compliance as we did previously (Figure 3.12). As we saw earlier, the lungs tend to collapse, so we need

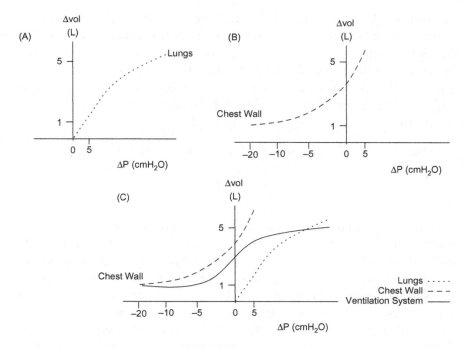

Figure 3.12 The compliance curve of the entire ventilation system (solid black line) (C) is a function of the combined compliance of the compliance of the lungs, which tend to collapse (A), and the compliance of the chest wall which tends to stay open (B).

a positive pressure to keep them open. We can see in Figure 3.12A that as pressure increases, the volume of the lungs increases. The chest wall functions in the opposite manner. The thoracic cavity tends to stay open at baseline, so we need to decrease the pressure inside the chest to get it to collapse. In Figure 3.12B, we see exactly that. As we decrease the pressure, the volume of the chest decreases. The compliance of our entire ventilation system, the lungs situated inside the thoracic cavity, is a combination of these two compliance curves, inward and outward. And that is exactly what we see in Figure 3.12C. The compliance to the ventilation system as whole (solid line) sits in the middle of both compliance curves. The take-home message from this graph is that altering the compliance of either chest wall or the lungs can alter whole ventilation system compliance.

Alveolar Diameter and Role of Surfactant

So far we've discussed "lung resistance" as an abstract concept that is generated by the tissue characteristics of the lungs themselves. It's time to go into a little more detail. Of the lung resistance we have been discussing, approximately one third is due to the collagen and elastin fibers that are woven into the lung tissue. The remaining two thirds, however, are the result of surface tension. In very simple terms, surface tension is the pull generated by the water molecules that are in contact with air. But in order to understand how surface tension works, let's take a step back and study the law of Laplace.

The law of Laplace states that for a spherical vessel:

$$P = \frac{2T}{r}$$

This means that the pressure to keep the alveoli open (P) is a function of the wall tension (T) and the radius (r). In the alveoli, the greatest contributor of wall tension is the surface tension of the water, which lines the alveolar wall. So, in practical terms, the amount of pressure required to keep the alveoli open (P) increases as the radius (r) decreases or the surface tension (T) increases. When discussing the lungs, P can be roughly equated to transpulmonary pressure. Therefore the transpulmonary pressure required to open the alveoli is a function of the surface tension and the radius of the alveoli. If the radius is small and/or the surface tension is high, we're going to need a higher transpulmonary pressure not only to keep the alveoli open, but also to move air in and out.

Key

A decreased radius or increased surface tension leads to decreased compliance and greater resistance to inflation.

Taking whole ventilation system compliance into account, let's understand the role that surface tension plays on compliance. Take a look at Figure 3.13. On the X-axis we will plot the pleural pressure and on the Y-axis we will plot the ΔVol. Let us draw your attention to the first curve (A). In the example in curve (A), the lungs are filled with saline solution. Even if the lungs are filled with fluid, pressure can still be applied to them in order to expand them. The dotted line (A) represents the compliance curve as the pressure is decreased and increased in saline-filled lungs. The solid line (B), on the other hand, represents the lungs that are filled with air. As amazing as it may

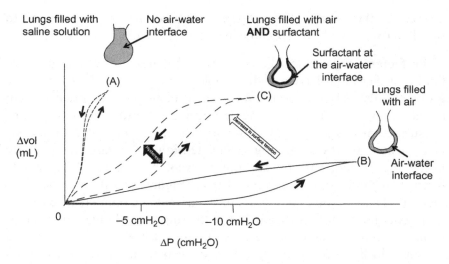

Figure 3.13 Whole ventilation system compliance curves for lungs filled with saline solution (A) no air water interface, lungs filled with air but with no surfactant; (B) high surface tension from air−water interface, and lungs filled with air AND surfactant; (C) decreased surface tension provided by the surfactant, which makes the lungs more compliant.

seem, these compliance curves represent the same lung tissue under different conditions. It only takes a very small amount of pressure to inflate the lungs in A, while the lungs in curve B, even though they're subjected to huge pressures, never really inflate to their full volume. So what's going on?

In Curve A, all the lungs are filled with saline solution, so there is no air−water interface at the level of the alveoli. In this way, only tissue resistance (i.e., collagen and elastin fibers) opposes inflation. With Curve B something else is going on. In physiologic conditions, there is always a small amount of water that lines the walls of all alveoli. (Remember from Chapter 2 that the partial pressure of H_2O in the airways is 47 mmHg.) When the lungs are filled with air, the air−water interface that forms at the level of the alveoli generates surface tension. As we said, surface tension is the pull generated by the water molecules that are in contact with air. As the water molecules "pull" toward one another the generated tension, which tends to collapse the alveoli, makes the radius smaller. And, according to the law of Laplace, the smaller the radius, the higher the pressure required to inflate the alveoli.

As you can see, the resistance that the air−water interface possesses is huge. (Just take a look at curve B!) So, how does the body counteract the enormous resistance generated by surface tension? In simple terms, it makes a special chemical agent called surfactant.

Surfactant, or "surface active agent" is the name give to a group of lipid molecules that have a hydrophobic group (tails) and a hydrophilic group (heads). This property allows them to position themselves right at the air—water interface (imagine pouring oil into a glass of water). As surfactant positions itself at the air—water interface, it disrupts surface tension, and by disrupting surface tension it decreases the pressure inside the alveoli, thereby curtailing alveolar collapse. What would happen to the compliance curve B from Figure 3.13 if we added surfactant? Take a look at compliance curve C. Compliance curve C represents lungs filled with air, but with surfactant lining the alveolar wall. The lungs in compliance curve C are a lot easier to inflate than the lungs in B, all due to the presence of surfactant. Surfactant will not only decrease surface tension across the alveoli, but it will do so in a way that the pressures are evenly distributed across the lung. This is an important feature if trying to maintain an even inflation throughout the lungs. (It's no good if one part of the lung is very well inflated while another is collapsed.)

Clinical Correlate

Respiratory Distress Syndrome
One of the biggest concerns when treating preterm babies i.e. (babies born before 37 weeks of pregnancy) is lung maturity. Surfactant starts being produced in sufficient amounts after 32 weeks. If the baby is born before that, the lack of adequate surfactant production will make breathing extremely difficult. If surfactant is not present, the surface tension will be very high and the pressures required to move air in and out are too high for a baby's tiny chest to be able to move air efficiently. This is called Respiratory Distress Syndrome (RDS). The lung compliance curves for lungs with RDS look similar to the curve C in Figure 3.13. The treatment for RDS is the administration of exogenous surfactant, which will help support the baby until he or she starts producing enough surfactant to allow for an adequate respiratory effort.

Of note, the pressure—volume loops shown in Figure 3.13 look somewhat different from the compliance graph we saw in Figure 3.12C. There seems to be a different compliance to the lungs as they're inflating versus. when they're deflating, no? This difference is termed *hysteresis*, which is the difference in compliance of the lungs in inspiration versus expiration. A simple, albeit incomplete explanation of this phenomenon is that as the lungs inflate, the alveolar radii increase. As each alveolar radius increases, the alveoli become more compliant (think Laplace). When the lungs are deflating, the opposite occurs. Since the radius is decreasing, the surface tension increases and it shifts the curve to the left. Therefore, even during normal quiet breathing there are variations in lung compliance.

As we have seen, ventilation system compliance can be altered by many factors including tissue resistance, surface tension, and chest wall dynamics. All these variables can alter the amount of air that moves in and out of the lungs in various pathological conditions, and this can ultimately have an impact on the underlying treatment of specific diseases.

CLINICAL VIGNETTES

Scenario 1

A previously healthy 30-year-old man comes into the Emergency Department after being involved in a motor vehicle accident (MVC). He was impacted on the left side by an oncoming vehicle. On arrival his vital signs are HR 99; RR 15; BP 110/65; O_2Sat − 98% on a 40% FiO_2 non-rebreather mask. He's complaining of chest pain, and there are decreased breath sounds on the left side. Since the patient is stable, a chest X-ray is ordered, which shows a large left-sided pneumothorax (air in the pleural space).

1. In this patient is the compliance of the left lung (pneumothorax) and the right lung the same?
   A. Left > Right
   B. Right > Left
   C. The same

Answer: B. Compliance as defined previously is the ΔVolume/ΔPressure; again in simpler terms, how much would the volume of each lung change if the same pressure was applied? In this case, the tissue resistance, which is provided by collagen and elastin, would be the same, but the total resistance is the summation of both tissue resistance and the resistance provided by wall tension. Since there is air in the left pleural space, the negative pressure required to keep the lung open has been lost. The loss of negative pressure then collapses the lung, which decreases the radius of all the alveoli. By decreasing the radius of the alveoli, the pressure required to inflate the left lung is now a lot greater than the pressure required to inflate the right lung.

2. What would be the best treatment for this patient?
   A. Endotracheal intubation with a very high positive pressure to inflate both lungs accordingly and improve the volume of the left lung.
   B. Endotracheal administration of surfactant to decrease the surface tension in the left lung so the patient can distend the left lung.
   C. Decompression of the left thoracic cavity with a chest tube in order to remove the air and recover the negative intrapleural pressure.
   D. Leave the patient alone, he seems to be able to breathe with one lung.

Answer: C. The goal of treatment is to try and correct the initial problem. In this case after the MVC the patient probably ruptured either the parietal or the visceral pleura, which allowed air to enter the pleural space. Even though the patient is stable, option D is not ideal given the mechanism of injury and the size of the pneumothorax. (Small pneumothoraces can be observed in a select group of patients, and the body will, slowly but surely, eliminate the excess air if there is no persistent leak.) Increasing the pressure of inflation through endotracheal intubation would not be ideal because both lungs would be subject to the increased pressure and this could potentially lead to an overdistention of one or both lungs, causing alveolar injury and further damage to the lung tissue. Administration of surfactant would also not be appropriate because assuming this patient was healthy, would be producing more than enough surfactant, adding more would have no effect on lung function.

Scenario 2
An obese 40-year-old man comes to your office for a check up. Since his last visit, he has put on 10 Kg, and now weighs 120 Kg and has a BMI of 44 kg/m². He complains of difficulty breathing when lying down, but after a thorough examination there is no indication of heart failure.

1. How would obesity impact the compliance of the ventilation system?
   A. Increase compliance
   B. Decrease compliance
   C. No impact on compliance

Answer: B. Obesity decreases chest wall compliance significantly. The addition of fat to the chest wall makes it more difficult to move. Additionally, the deposition of fat in the intercostal muscles, the abdominal cavity and the diaphragm decreases ventilation muscle function. This results in having an increased respiratory effort. (Remember decreased compliance means less volume for the same amount of pressure difference.)

CHAPTER 4

The Respiratory Cycle

In the previous chapter we studied the mechanics of how air moves in and out of the lungs. Now we'll take a closer look at the respiratory cycle. Clinically it would be great to know the intrapleural pressure and the alveolar pressures to aid the diagnosis and management of patients. However the alveolar and pleural pressures are not easy measurements to get! Therefore we must seek alternate parameters of clinical well being to get an idea of how lungs are working and through which we can monitor patients. This is where lung volumes and capacities step in. Throughout this chapter we will understand what each volume and capacity represents and how it impacts lung function in both normal and diseased states.

WHAT ARE LUNG VOLUMES AND CAPACITIES?

The amount of air going in and out of the lungs can be measured at any point in time using a device called a spirometer. The spirometer is attached to someone's mouth (assuming they won't breathe through their nose), and measures both the quantity and the flow of air in different phases of the respiratory cycle. From all the data that modern spirometers can give us, we're going to focus on only a couple of measurements:

- **Lung Volumes**. Amounts of air moving in and out of the lungs that can be measured directly or indirectly.
- **Lung Capacities**. Amounts of air moving in and out of the lungs that are composed of two or more lung volumes.

Figure 4.1 is a visual representation of both lung volumes and lung capacities. Lung volumes are represented by dotted arrows and capacities are represented by solid lines.

Back to Basics in Physiology. DOI: http://dx.doi.org/10.1016/B978-0-12-801768-5.00004-6

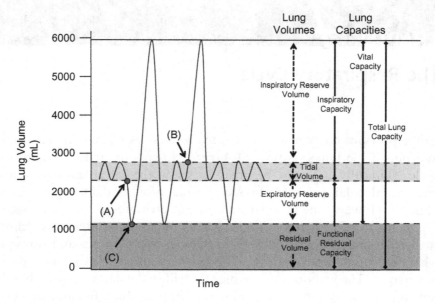

Figure 4.1 Lung volumes and capacities plotted in volume against time.

There are four volumes:

- Tidal Volume (V_T)
- Expiratory Reserve Volume (ERV)
- Inspiratory Reserve Volume (IRV)
- Residual Volume (RV)

and four capacities:

- Inspiratory Capacity (IC) = IRV + V_T
- Functional Residual Capacity (FRC) = ERV + RV
- Vital Capacity (VC) = IRV + V_T + ERV
- Total Lung Capacity (TLC) = IRV + V_T + ERV + RV

Let's have a go at the volumes first. Tidal volume (V_T; narrow greyed box in the middle of Figure 4.1) is the amount of air that moves in and out of the lungs during normal quiet breathing. It is approximately 500 mL, although this will vary from person to person according to chest and lung size. Notice that at the end of a cycle of normal quiet breathing (point A), once you're done exhaling, if you try you can still exhale even more. This volume is called the Expiratory Reserve Volume (ERV) and it's the amount of air that is in the lungs after normal quiet breathing. Similarly, at the inspiratory peak of normal quiet breathing (point B) there is still a large volume of air that can be inhaled. This is called Inspiratory Reserve Volume (IRV), and it's the amount of air that can still be brought into the lungs after normal quiet breathing. Residual Volume (RV), represented by a dark grey box at the bottom of Figure 4.1, is the amount of volume that cannot be exhaled and is always trapped in the lungs. Why is this?

If you think back to the compliance curve for the entire ventilation system (we'll redraw it for you in Figure 4.2), the chest wall opposes collapse while the lungs oppose expansion. The resulting curve is the compliance curve for the entire ventilation system (solid black line). Because the lungs sit in the chest and the chest can't completely collapse to a volume of 0 (no matter how much we decrease the pressure), the volume of air that will necessarily remain in the lungs is the RV (bottom grey box in Figure 4.2). The only theoretical way of getting rid of the residual volume is to make the lungs independent of the chest wall. And the only way to achieve this is with a pneumothorax. If the negative intrapleural pressure is lost (as would happen if you poke a hole in the chest wall), then the outward pull of the chest wall on the lungs will be lost and the lungs will shift to the lung compliance curve (dotted line in Figure 4.2) and collapse, thereby forcing any RV out. RV, unlike the other volumes, can't be measured directly because there's no way to get it out of the lungs (and generating bilateral pneumothoraces to force all the air out while our patient is connected to the spirometer is not exactly ethical).

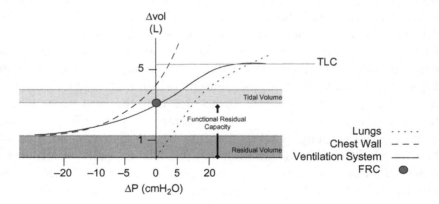

Figure 4.2 Compliance curve of the ventilation system where Total Lung Capacity (TLC) is represented by the top-most grey box; Functional Residual Capacity (FRC) is represented by the grey circle; and Residual Volume (RV) is indicated by a grey box. The point at which the ventilation system compliance curve crosses the Y-axis represents the zero point in the respiratory cycle; i.e., the moment when the outward elastic recoil of the chest wall balances out the inward elastic recoil of the lungs.

RV is measured through a helium dilutional technique. Think about it like this: If you have a an unknown quantity of air in the lungs you can estimate how much air is in there by adding a known volume of air that will mix with the unknown quantity. Use the following equation to calculate the result: $C_1 \times V_1 = C_2 \times V_2$ where C = concentration and V = volume. Essentially, you connect the patient to the spirometer at point C in Figure 4.1 and make him or her breathe in helium. As the patient breathes in, the helium will mix with the RV. Then you ask the patient to breathe out and the RV can be calculated from the concentration of helium in the expired air.

Key

Tidal Volume (V_T) is the amount of air that moves in and out of the lungs during a passive respiratory cycle. It is approximately 500 mL.

That explains the volumes, but how about lung capacities? As we said earlier, capacities are functions of lung volumes. Inspiratory Capacity (IC) is the amount of air that can be inhaled after passive expiration. It is the sum of the IRV and the V_T. Vital Capacity (VC) is the amount of air that can move in and out of the lungs. It is the sum of the IRV, the ERV, and the V_T. Total lung capacity is all the air that can be present in the lungs at any given point in time and it is the sum of all lung volumes. The last capacity is called the Functional Residual Capacity (FRC). It is the sum of the RV and the ERV, and it is called "the lung's physiologic reserve." FRC is the point where the outward expansion of the chest wall balances out with the lungs' tendency to collapse (black circle in Figure 4.2). Therefore it is the amount of air that is left inside the lungs after passive expiration. In essence it is the zero point where the respiratory cycle starts. It is important to keep FRC in mind while evaluating lung function because it will give you an idea of how much effort the ventilation system has to exert in order to move air in and out. If there are changes in ventilation system compliance, the FRC will be modified and this will impact not only the exertion required to move air in and out of the lungs, but the residual amount of air that is sitting in the alveoli and can participate in gas exchange.

Clinical Correlate

Functional Residual Capacity (FRC)

Pulmonary diseases can both increase and decrease the FRC. Diseases that increase the FRC are called obstructive lung disease, because they obstruct the exit of air from the lungs, and cause air trapping. Examples are COPD and asthma. Diseases that decrease the FRC are called restrictive lung diseases because they decrease the compliance of the ventilation system and therefore make the movement of air in and out of the lungs more difficult. Examples are interstitial lung diseases, muscle diseases that paralyze respiratory muscle function, or diseases that affect chest wall function such as obesity or kyphoscoliosis. In both instances, the work required to move air in and out of the lungs is increased, thereby leading to problems with either providing O_2, removing CO_2, or both.

ALVEOLAR VENTILATION AND DEAD SPACE VENTILATION

Remember that the overall goal to lung function is to move air in and out to provide O_2 and remove CO_2 from the body. In order to do this, the air in the areas where gas exchange occurs needs to be renewed constantly. However, not all the air that is inhaled participates in gas exchange. The total amount of air that moves in and out of the ventilation system is called minute ventilation.

$$\text{Minute Ventilation} = RR \times V_T$$

where:

Minute Ventilation = Total amount of air going into the ventilation system

RR (Respiratory Rate) = Number of breaths we take in a minute (let's say a normal RR is 12)

V_T (Tidal volume) = Air that is inspired during a respiratory cycle (in standard conditions we say it's approximately 500 mL)

So substituting:

$$\text{Minute Ventilation} = 12 \times 500 \text{ mL}$$

this would come out to a minute ventilation of 6000 mL of air per minute!

But this is the minute ventilation, and not all the air we breathe in reaches the alveoli. Therefore in order to calculate how much air is actually reaching the alveoli and participating in exchange (known as Alveolar Ventilation, or V_A) we need to calculate the volume of air that is trapped in areas of the lung where exchange is not taking place and subtract that from minute ventilation.

Areas of the lung where exchange cannot take place are called dead space. There are two types of dead space:

- **Anatomic dead space**. Areas of the lung where exchange simply cannot take place because they are not designed for it. Anatomic dead space is made up of the conducting airways (trachea, bronchi, etc.). It is approximately 150 mL.
- **Physiologic dead space**. Areas of the lung where exchange could take place, but where it is not occurring due to lack of blood flow to a particular alveolar segment. Under normal conditions, physiologic dead space is minimal but conditions that decrease alveolar blood flow can increase the physiologic dead space to levels that are incompatible with life.

Key

Alveolar ventilation is the volume of inspired air that reaches the alveoli and participates in gas exchange.

Clinical Correlate

Pulmonary Embolism and Dead Space

A pulmonary embolism is the lodging of a blood clot that was formed somewhere else inside the body (generally the veins in the lower extremities), travels through the venous system, and gets lodged in the pulmonary vasculature. This is a medical emergency because the decreased blood flow to the alveoli decreases the surface area available for exchange. As the blood clot lodges in a branch of the pulmonary artery, all the alveoli that were perfused by that segment now become dead space ventilation. This decrease in gas exchange, combined with the increased resistance in the pulmonary blood vessels, can overwhelm the right ventricle (which is used to dealing with low pressures), leading to cardiac insufficiency and potentially death.

We can calculate alveolar ventilation using the formula:

$$V_A = RR \times (V_T - V_D)$$

where:

V_A (Alveolar Ventilation) = The amount of air in (mL/min) that is participating in exchange
RR (Respiratory Rate) = The number of breaths we take in a minute
V_T (Tidal volume) = The air that is inspired during a respiratory cycle; as stated previously, in standard conditions it's approximately 500 mL
V_D (Dead Space) = The amount of air that is inspired per breath and is *not* participating in exchange

Under standard conditions:

$V_T = 500$ mL
$V_D = 150$ mL
$RR = 12$ in 1 minute

Therefore:

$$V_A = 12 \times (500 - 150)$$

or

$$V_A = 4,200 \text{ mL/min}$$

This means that even though the total amount of air being inspired (minute ventilation) is 6000 mL, only 4200 mL actually participate in gas exchange.

Alveolar ventilation is one of the most important concepts in ventilation physiology. It is one of the major determining factors of O_2 and CO_2 exchange, so let's spend a little more time trying to fully understand its determinants.

During periods of increased O_2 consumption and CO_2 production, such as aerobic exercise, alveolar ventilation must increase in order to fit the demand. In what ways can we increase alveolar ventilation most effectively? Well, we can try and modify V_D, RR, or V_T.

- **Dead Space (V_D).** Under nondiseased conditions dead space is mainly anatomic dead space, which cannot be modified. In diseased states where there's an increase in physiologic dead space, attempts to improve exchange in the nonfunctioning alveoli is critical. So, it's important to consider that in certain patients there might be pathologic states associated with increases in physiologic dead space that could impact their ability to provide O_2 and clear CO_2 from the body.

- **Respiratory Rate (RR).** If we increase respiratory rate, we are sure to increase V_A. But, how much? Let's say we double our RR from 12 per minute to 24 breaths per minute; how will this modify V_A? In our previous example:

$$V_A = 12 \times (500 - 150) = 4,200 \text{ mL/min}$$

If we double the RR then:

$$V_A = 24 \times (500 - 150) = 8,400 \text{ mL/min}$$

Hence doubling of the RR leads to an exact doubling of the V_A. But this is not sustainable over the long term due to the amount of energy required to breathe so many times so fast! In fact, in some patients with a very high RR, intubation is considered a way of protecting the airway because patients will tire out, and as soon as they tire out, they will be unable to meet their O_2 needs.

- **Tidal Volume (V_T).** Similar to what happens with RR, increasing V_T will surely increase V_A, but by how much? Let's double V_T and see what happens. From our previous example:

$$V_A = 12 \times (500 - 150) = 4,200 \text{ mL/min}$$

If we double V_T then:

$$V_A = 12 \times (1000 - 150) = 10,200 \text{ mL/min}$$

This represents an increase in V_A that is 20% greater than that if we increase RR, so increasing V_T is more effective than increasing RR. But taking such deep breaths is also extremely tiring, which is why the body doesn't alter only a single variable at a time.

In reality, when we need to increase V_A, the body attempts to increase both V_T and RR simultaneously. This is the most effective way to increase V_A. If instead of doubling the RR or the V_T we increase each by half, what do you think will happen to V_A?

From our previous example:

$$V_A = 12 \times (500 - 150) = 4,200 \text{ mL/min}$$

If we increase V_T and RR by 50% each, then:

$$V_A = 18 \times (750 - 150) = 10,800 \text{ mL/min}$$

This represents an even bigger increase than what we saw by doubling RR or V_T alone!

COMPOSITION OF ALVEOLAR AIR

Alveolar air is a product of both alveolar ventilation, as we studied in the previous section, and the rate at which blood extracts O_2 from and releases CO_2 into the alveoli. However, keep in mind that in spite of alveolar ventilation being approximately 4,200 mL/min under steady state conditions, the turnover rate of alveolar air is surprisingly small. What this means is that the complete renewal of the air that is present in the alveoli does not happen with each breath. The main reason for this is the FRC. The FRC is 2300 mL, and with each breath approximately 350 mL of air gets inspired and expired. This means that only 15% of the FRC is exchanged with each breath. Think of this as a failsafe of sorts. A small turnover rate allows for a gradual gas exchange, which means that even though we are breathing in and out all the time, there are no sudden changes to blood gas concentrations.

So, what exactly are the partial pressures of O_2 and CO_2 in the alveoli? Table 4.1 shows the progressive changes in PO_2 and PCO_2 as air moves down the respiratory system. In a nutshell PO_2 decreases and PCO_2 increases the closer we get to the alveoli. This should be intuitive as O_2 is being consumed and CO_2 is being produced. The initial decrease in O_2 from atmospheric to moist tracheal air is due to the increase in PH_2O from 0 to 47 mmHg. Then PO_2 continues to decrease

Table 4.1 Approximate Standard Partial Pressures of Gases at Sea Level on an Average Day				
Partial Pressures of Gases in mmHg				
	Atmospheric Air	Moist Tracheal Air	Alveolar Air	Expired Air
PO_2	160	150	104	120
PCO_2	0	0	40	27
PH_2O	0	47	47	47
PN_2	600	563	569	566
Ptotal	760	760	760	760

and CO_2 continues to increase until the air reaches the alveoli. However, in order for exchange to take place appropriately the alveolar pressure of O_2 or P_AO_2 should be approximately 100 mmHg at sea level, and the alveolar pressure of CO_2 or P_ACO_2 should be approximately 40 mmHg. Commit these numbers to memory.

Key

At sea level the alveolar pressure of O_2 (P_AO_2) is approximately 100 mmHg, and the alveolar pressure of CO_2 (P_ACO_2) is approximately 40 mmHg.

Clinical Correlate

FiO$_2$: Fraction of inspired oxygen
As we said, 21% of the atmospheric air is O_2. Therefore the fraction of air we inspire that is O_2 or (FiO$_2$) is usually 21%. However, when a patient is sick from a respiratory illness (or even from anemia or shock), giving O_2 is an imperative part of stabilizing the patient. A simple way to get *more* O_2 to a patient is by increasing the amount of O_2 inhaled with each breath. Thus the blood will pick up more O_2 and will end up delivering more O_2 to cells in the body. Therefore, increasing the partial pressure of O_2 by providing supplemental O_2, and thus altering the FiO$_2$, can be critical in unstable patients.

The P_AO_2 and the P_ACO_2

Remember, the goal is to get O_2 into the body! To do this, we need to increase the amount of O_2 in the alveoli. So let's take a look at two factors that regulate the alveolar pressure of O_2 (P_AO_2):

- **The rate at which O_2 is brought in by the ventilation system**. In broad terms this means that the higher the alveolar ventilation (V_A) the more the P_AO_2 is going to approximate the pressure of O_2 in the atmospheric air.
- **The rate of O_2 extraction by the pulmonary capillaries**. In a nondiseased state the amount of O_2 that the pulmonary capillaries extract from the alveoli is directly related to the amount of O_2 being consumed by the body. This means that the body's metabolic rate is directly related to the pulmonary extraction of O_2.

The relationship between P_AO_2, alveolar ventilation, and O_2 consumption is represented in Figure 4.3A. Alveolar ventilation in

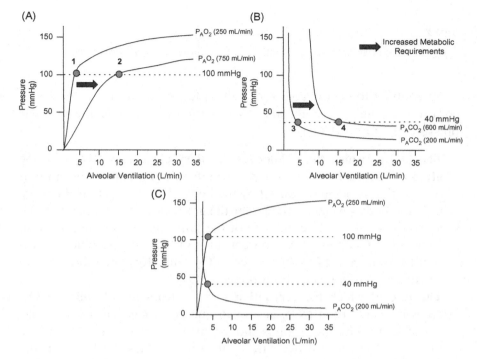

Figure 4.3 Relationship between alveolar ventilation (V_A), P_AO_2 (A), and P_ACO_2 (B), and both combined (C). An increase in metabolic requirements (block arrow) leads to an increased consumption of O_2 and an increased production of CO_2. This in turn shifts both curves to the right, requiring an increase in alveolar ventilation from 4.2 L to approximately 15 L to reach normal levels of P_AO_2 and P_ACO_2 once again.

L/min is on the X-axis, while pressure of O_2 in mmHg is on the Y-axis. The top curve represents a standard O_2 consumption of 250 mL/min. Maintaining a P_AO_2 of approximately 100 mmHg requires a V_A of approximately 5 L (point 1). If we were to maintain O_2 consumption stable at 250 mL/min and we changed the V_A, the P_AO_2 would follow V_A; that is, as V_A increases the P_AO_2 would increase and if V_A decreases the P_AO_2 would decrease. However in the setting of increased metabolic requirements (e.g., aerobic exercise), there is an increase in O_2 consumption. Imagine if we run to catch the bus and this tripled O_2 consumption of O_2 from 250 mL/min to 750 mL/min (bottom curve). If V_A stayed at 5 L/min, the P_AO_2 would be around 50 mmHg!! So in order to return the P_AO_2 to 100 mmHg, V_A must also triple and increase from around 5 L/min to 15 L/min (point 2).

Key

P_A denotes *alveolar* pressures whereas P_a denotes *arterial* pressures.

In contrast to O_2 the goal with CO_2 is to get it out of the body, and curiously enough there are also two factor that regulate the alveolar pressure of CO_2 (P_ACO_2):

- **The rate at which the pulmonary capillary exchanges CO_2 with the alveoli.** Essentially in a non-diseased state this amounts to the amount of CO_2 that is being produced by the body. This means that the more CO_2 the body produces, the larger the gradient for exchange with the alveolus. This is a different side of the same energy production coin. Remember Chapter 1? As you consume O_2 through aerobic respiration to produce ATP, you produce CO_2. Therefore the more O_2 you consume the more CO_2 you produce.
- **The rate at which the ventilation system clears the alveoli of CO_2.** Alveolar ventilation (V_A), is also in charge of clearing the alveolar air of CO_2. The more air is moved in and out of the alveoli, the more the P_ACO_2 will decrease as it approximates the PCO_2 in the atmospheric air, which is close to zero. This means that the higher the rate of alveolar ventilation (V_A), the lower the pressure of CO_2 at the alveoli level is going to be. The relationship between alveolar ventilation and arterial pressure of CO_2 or $PaCO_2$ is summarized in the alveolar CO_2 equation or P_ACO_2 equation, which states that:

$$P_aCO_2 = \frac{VCO_2}{V_A}$$

where:

- P_aCO_2 is the pressure of CO_2 in the arterial blood.
- VCO_2 is the amount of CO_2 produced by the body and delivered to the lungs.
- V_A is alveolar ventilation.

This relationship is based on the idea that the pressure of CO_2 in the alveolus (P_ACO_2) and the pressure of CO_2 in arterial blood (P_aCO_2) is essentially equivalent (remember CO_2 diffuses extremely fast). The take-home message from this equation is that if V_A decreases the P_aCO_2 will increase and if V_A increases the P_aCO_2 will decrease.

Key

If V_A decreases the P_aCO_2 will increase. If V_A increases the P_aCO_2 will decrease.

This relationship is graphed in Figure 4.3B. The axis are the same as those in Figure 4.3A, but this time the curves represent P_ACO_2. The bottom-most curve represents a normal production of 200 mL of CO_2 per minute. At this rate of CO_2 production V_A needs to be around 5 L/min to maintain a P_ACO_2 of 40 mmHg. As defined by the P_aCO_2 equation, if the production of CO_2 remains stable at 200 mL/min, increases in V_A will decrease the P_ACO_2, while a decrease in V_A will increase CO_2. If we were to increase the production of CO_2 from 200 mL/min to 600 mL/min because we ran to catch the bus (top curve), we would have to increase V_A from 5 L/min to 15 L/min in order to maintain a P_ACO_2 of 40 mmHg.

The body, being the amazing feat of biological engineering that it is, tries to make efficient use of all its biological processes. Therefore the same alveolar ventilation that brings in O_2 is in charge of clearing CO_2! This is represented in Figure 4.3C, where we have superimposed Figures 4.3A and 4.3B, and you can see how P_AO_2, P_ACO_2, and V_A are interrelated. Increases in V_A will result in both an increase in the P_AO_2 and a decrease in P_ACO_2, while a decrease in V_A will result in a decrease in P_AO_2 and an increase in P_ACO_2. Think about it like this: If you were to hold your breath right now, you would decrease V_A, and by doing so the P_AO_2 would decrease and the P_ACO_2 would increase. (If you're hypoventilating, you're still consuming O_2 without bringing any new O_2 in, and you're still producing CO_2 without dumping any of it out into the atmosphere!) Conversely if you were to increase V_A by hyperventilating, the P_AO_2 would increase and the P_ACO_2 would decrease. (If you're hyperventilating, you're bringing in more O_2 than is being consumed, and dumping out more CO_2 than is being produced.)

However, the increase in P_AO_2 is not of the same magnitude as the decrease in P_ACO_2. Why? Well, for starters take a look at the amount of CO_2 that is produced (200 mL) and the amount of O_2 that is consumed (250 mL). Not exactly a 1:1 ratio is it? The relationship between O_2 that is consumed to the CO_2 that is produced is called the Respiratory Quotient (RQ). The RQ will depend on the molar ratios of O_2 consumption to CO_2 production, from the substrate that is being used as fuel by the body. When we use glucose the ratio is 1:1; that is, you consume one mole O_2 for every mole CO_2 you produce. However the body also uses fat and protein as fuel, which aren't as efficient. The RQ in the body approximates 0.8 (200 mL CO_2/250 mL O_2). In other words, for every mole of O_2 consumed, the body produces approximately 0.8 moles of CO_2. Therefore the RQ helps us estimate how much O_2 we consumed in order to produce a given amount of CO_2. Why is this important? Well, in clinical practice it's extremely difficult to directly measure the P_AO_2, so we calculate it using something called the Alveolar Gas equation. To do so we need to understand the RQ.

The Alveolar Gas Equation, or, How Much O_2 is in There!

In the previous section we stated that the two factors that regulated the P_AO_2 were the renewal of O_2 through ventilation and the rate of consumption. Since it is relatively difficult to measure exactly how much O_2 is in the alveoli, wouldn't it be nice if we could put this in a formula to calculate the P_AO_2? Well, there is a formula and it's called the alveolar gas equation; it looks something like this:

$$P_AO_2 = FiO_2(P_{ATM} - P_{H_2O}) - \frac{P_aCO_2}{RQ}$$

However, before we delve into the specifics of each variable let's take a step back. Consider this: If we're trying to calculate the P_AO_2 we need to know two things:

- How much O_2 is being inspired
- How much O_2 is being consumed

Therefore, here's a simplified version of the alveolar gas equation:

$$P_AO_2 = O_2 \text{ inspired} - O_2 \text{ consumed}$$

Since most of the time we are not breathing pure O_2, we'll need to calculate the partial pressure of O_2. We can calculate how much O_2 is being inspired from partial pressure of O_2 in the air that is being breathed in (in this case we'll use atmospheric air) and the PH_2O in the airways with the following formula:

$$\text{Inspired } O_2 = FiO_2(P_{ATM} - P_{H_2O})$$

where:

P_{ATM} = Atmospheric pressure (at sea level it would be 760 mmHg).
FiO_2 (Fraction of Inspired Oxygen) = The percentage partial pressure of O_2 in the air that the patient is breathing in. Since O_2 makes up 21% of the air we breathe under normal conditions the FiO_2 would be 0.21 if no extra oxygen is added. (If O_2 is added to the mix by providing supplemental oxygen to the patient the FiO_2 will increase.)
PH_2O = Partial pressure of H_2O in the system. At normal body temperature this is equal to 47 mmHg.

If we plug the numbers into our formula we come up with the following:

$$\text{Inspired } O_2 = 0.21(760 \text{ mmHg} - 47 \text{ mmHg})$$

Then:

$$\text{Inspired } O_2 = 0.21(713 \text{ mmHg})$$

So:

$$\text{Inspired } O_2 = 150 \text{ mmHg}$$

This number is the same number we found in Table 4.1 for moistened tracheal air. So far, so good! Now our simplified alveolar gas equation looks like this:

$$P_A O_2 = 150 \text{ mmHg} - O_2 \text{ consumed}$$

How do we calculate O_2 consumption? Remember what we mentioned about RQ in the previous section? RQ is the ratio of CO_2 produced to O_2 consumed. So if we know the $P_A CO_2$ we can estimate how much O_2 is being consumed. Lucky for us the CO_2 in arterial blood or $P_a CO_2$ is almost identical to the $P_A CO_2$. CO_2 diffuses extremely rapidly, and is generally not affected by issues that can alter O_2 diffusion (keep in mind that CO_2 diffuses about 20 times as fast as O_2). Therefore if $PaCO_2 \approx P_A CO_2$ together with the RQ, we can then calculate how much O_2 is being consumed:

$$O_2 \text{ consumed} = \frac{P_a CO_2}{RQ}$$

where:

P_aCO_2 = The alveolar pressure of CO_2, which can be inferred from an arterial blood sample. Assuming that diffusion is occurring unimpinged, the amount of CO_2 in the blood, the $PaCO_2$ is going to be very similar to the P_ACO_2, and is therefore used as a surrogate.

RQ = The molar ratio of CO_2 produced to O_2 consumed depending on the fuel being consumed (carbohydrates vs fats vs proteins). In the body it approximates 0.8 and is product of the combined metabolism of carbohydrates, fats, and proteins.

Key

The RQ is the amount of O_2 in mmHg that has to be consumed to account for the amount of CO_2 in mmHg that is being produced.

Now that we understand the components let's look at the entire alveolar gas equation:

$$P_AO_2 = O_2 \text{ inspired} - O_2 \text{ consumed}$$

or

$$P_AO_2 = FiO_2(P_{ATM} - P_{H_2O}) - \frac{P_aCO_2}{RQ}$$

So, let's plug in the numbers under standard conditions and see what we come up with.

$$P_AO_2 = 0.21(760 - 47) - \frac{40}{0.8}$$

As we saw previously, if we solve the first half first we come up with 150 mmHg.

$$P_AO_2 = 150 - \frac{40}{0.8}$$

Dividing by 40 by 0.8 yields 50 mmHg. Let's go over this again. This number is telling us that for every 40 mmHg of CO_2 that we produce we are consuming 50 mmHg of O_2. So if your P_ACO_2 is 40, it's because we are consuming 50 mmHg of O_2. Therefore, we subtract the amount of O_2 that is being consumed (50 mmHg) from the amount that is being brought into the lungs (150 mmHg). So:

$$P_AO_2 = 150 \text{ mmHg} - 50 \text{ mmHg}$$

Then:

$$P_AO_2 = 100 \text{ mmHg}$$

Success! The P_AO_2 we calculated is right on the money! Taking small variations into account, the normal P_AO_2 is approximately 100 mmHg.

Key

The alveolar gas equation is a way to estimate the P_AO_2, and is calculated with the following formula: $P_AO_2 = FiO_2 \ (P_{ATM} - P_{H_2O}) - P_aCO_2/RQ$.

Knowing the pressures of O_2 and CO_2 at the level of the alveolus, however, is only half the battle! We now need to understand what happens in the blood that allows for the diffusion of these gases from the air and into the blood, and how hemoglobin allows this phenomenon to take place in quantities that are compatible with life.

CLINICAL VIGNETTES

A 45-year-old male comes in to the Emergency Department after being found unconscious at his beach house. On arrival his vital signs are HR 60, BP 95/55, RR 2, Temp 36.2°C, and O_2 Sat 70%. On quick examination he is unarousable, has pinpoint pupils, and you note recent needle sticks in his left forearm. The patient is intubated with an FiO_2 of 40% and a respiratory rate of 12 breaths per minute, his saturation rapidly increases to 95%, his $PaCO_2$ is 60 mmHg, and his PaO_2 is 160 mmHg.

1. What is the most likely diagnosis?
    A. The patient is suffering from an opioid overdose.
    B. The patient was holding his breath and passed out.
    C. The patient had a seizure and is now in a post-ictal state.

Answer: A. Given the history of being found unconscious with pin-point pupils and decreased respiratory rate, we must always think of an opiate overdose. The needle sticks in his left forearm support the presumptive diagnosis of opiate intoxication. Answer **B** is wrong, because even if he were capable of holding his breath until he passed out, once he was out, his respiratory drive would kick in and he would begin breathing again. After a seizure patients can have a post-ictal period (which literally means after the seizure or ictal period). The post-ictal period can be characterized by an altered level of consciousness, confusion, and various neurological abnormalities including paralysis of the brain region affected by the seizure. Patients in the post-ictal state can have vomiting and aspiration, which would impair their ventilation. In this case, although aspiration can't be ruled out, the most likely diagnosis is an opiate overdose.

2. What would the calculated P_AO_2 of this patient be?
   A. 50 mmHg
   B. 86 mmHg
   C. 100 mmHg
   D. 185 mmHg

Answer: D. We would need to use the alveolar gas equation to calculate the estimated P_AO_2. The FiO_2 is 40%. This means that at sea level (he was found at his beach house) the pressure of O_2 in the moistened airway is $= FiO_2(P_{ATM} - P_{H_2O})$ or $0.4(760 - 47) = 285$ mmHg, and 80 mmHg $CO_2/0.8 = 100$ mmHg, therefore 285 mmHg $-$ 100 mmHg yields a calculated P_AO_2 of 185 mmHg.

3. Prior to intubation, what would the V_A of this patient be if his VT is 200 mL?
   A. 1000 mL/min
   B. 350 mL/min
   C. 700 mL/min
   D. 100 mL/min

Answer: D. Since this patient is overdosed on opiates, his respiratory effort is almost completely depressed. This means that both his respiratory rate and depth of inspiration are going to be impaired. This is exactly the case when the question states that his V_T is only 200 mL. This is a *very* shallow breath if we use the RR provided by the stem of the question (2 breaths per min). Assuming dead space

(V_D) is standard and there's no other underlying pathology, our V_A equation states that:

$$V_A = RR(V_T - V_D)$$

Therefore:

$$V_A = 2(200 - 150) \text{ or } 100 \text{ mL/min}$$

It should go without mentioning that this is completely inadequate, and finding an elevated $PaCO_2$ should not be a surprise. In the setting of such a decreased V_A, the increased $PaCO_2$ should be interpreted as a decreased exchange rather than an increased production of CO_2.

CHAPTER 5

Gases Inside the Body, Liquid Transport

Once atmospheric air makes its way down to the alveolus, it next has to overcome the huge hurdle of making its way into the blood. This is not a trivial step. O_2 and CO_2 need to move from a gas phase (alveolar air) to a fluid phase (blood) before they can move throughout the body in the cardiovascular system. This is a tough problem because diffusion in the gas phase is a lot easier than diffusion in the liquid phase. Why? Think about it: liquid is far more compressed/dense than gas, which means that the speed at which diffusion of gases in liquid phase takes place is thousands of times slower than the speed of diffusion in gas phase. (It's a lot harder to get where you're going if there are more obstacles in your way.) Bottom line, getting gases into a liquid is hard.

GETTING INTO BLOOD, HENRY'S LAW, AND WHY WE NEED RED BLOOD CELLS

How well a particular gas dissolves in a liquid will help determine the amount of that gas in a liquid at a given pressure. Wait, what? Well, this is the concept behind Henry's law. Henry's law tells us that the amount of dissolved gas that we are going to find in a liquid is dependent on the partial pressure of the gas and the solubility of the gas in that liquid. Ok, sure, that may be true getting O_2 and CO_2 to diffuse into water, but blood is special, isn't it? Isn't it? Well, if we're talking about blood *plasma*, it isn't. In order to understand exactly how much of a gas is in any liquid we need to think about its solubility. It turns out that the solubility of CO_2 in plasma is low, but the solubility of O_2 in plasma is really, really, really low! So much so, that at the pressures of O_2 that exist in the body (remember from the last chapter that the alveolar pressure of O_2 or P_AO_2 is approximately 100 mmHg), the amount of O_2 that is going to be dissolved in plasma would be around 15 mL of O_2 for the entire circulation. This is close to nothing when compared to the baseline consumption of 250 mL of O_2 per minute! This would leave us about 235 mL of O_2 short every minute. So there's got to be a better way to do this...

Back to Basics in Physiology. DOI: http://dx.doi.org/10.1016/B978-0-12-801768-5.00005-8

Well, the body came up with two particularly genius innovations to solve the problem of moving O_2 and CO_2 around: blood and the cardiovascular system.

- Blood holds on to the dissolved and nondissolved O_2 and CO_2.
- The cardiovascular system moves the blood around the body.

Think about it like this: O_2 and CO_2 in the body are packages that need to be delivered by the postal service. Now, think about how the postal service works.

Essentially, mail trucks go to the shipping centers where they drop off their outgoing packages and pick up incoming packages, after which they head out, first at high speed through the interstates then through progressively smaller roads at a lower speed. Finally, the mail trucks reach their intended destination where they drop of the packages to be delivered and pick up the outgoing packages and then head back to the shipping center to repeat the cycle all over again.

If we exchange a couple of words from the previous paragraph, this analogy works for O_2 and CO_2!

Essentially, *red blood cells* (*RBCs*) go to the *lungs* where they drop off *CO_2* and pick up *O_2*. Then they head out, first at high speed through the *arteries* then through progressively smaller *arteries* and *arterioles* at a lower speed. Finally, the blood reaches the *systemic capillaries* where it drops off O_2 and picks up CO_2 and then heads back to the lungs to repeat the cycle all over again.

The previous paragraph is a very succinct explanation of what blood does in the body. But, in order to understand how exactly blood does what it does, first we need to understand what it is. Blood is combination of water, salts, other solutes and cells. We can basically divide blood into two major parts: (1) RBCs, also known as erythrocytes, and (2) blood plasma (Figure 5.1A). Plasma (the liquid part of blood) usually is around 55% of the total blood volume. It is made up of about 92% water, 7% vital proteins such as albumin, and things like clotting factors, fats, sugars, vitamins, and salts. The remaining 45% of the total blood volume is made up of RBCs, and less than 1% are white blood cells and platelets that have no bearing on oxygen delivery but are very important in fighting infections and clotting (Figure 5.1B).

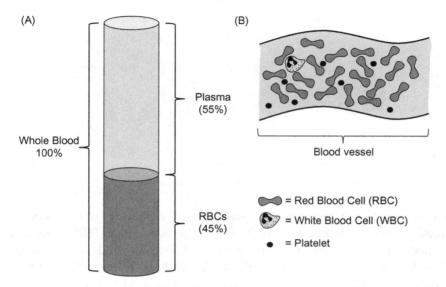

Figure 5.1 Whole blood can be divided into plasma and RBCs (A). The fraction of plasma that is composed of RBCs is known as the hematocrit. Among the other cellular components of blood are white blood cells (WBCs) and platelets (B).

WHY ARE RBCs SO SPECIAL?

As we said previously O_2 can't readily dissolve in plasma, so the body needs another way to move O_2 around. So, let's take a closer look at our delivery trucks. What makes RBCs different? They contain an almost magical substance called hemoglobin. Hemoglobin is what makes RBCs specialized carriers of oxygen. Hemoglobin reversibly binds with both O_2 and CO_2 increasing the blood's CO_2 and O_2 carrying capacity by several orders of magnitude (more on this later).

But a good delivery truck is only as good as its ability to get to the intended destination. Can you imagine an 18-wheeler trying to navigate cul de sacs to deliver Valentine's Day cards? A truck like that would probably get stuck at some point as it tries to navigate residential streets. So, you need a smaller more nimble truck to navigate the small streets. How is this related to RBCs? Well, capillaries are extremely thin, in fact, so thin that RBCs are about 25% *larger* than the capillaries. So, how do the RBCs manage to get around? Well, unlike other cells in the body, their shape is that of a biconcave disc. The classic description is that under the microscope RBCs have "a central area of pallor" due to "excess" membrane (Figure 5.2). All that

Side View   Top View

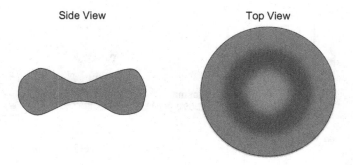

Figure 5.2 The shape of a red blood cell. Note the central area of pallor in the top view.

extra membrane increases surface area of exchange and makes the RBCs flexible, which is required for the job, since RBCs need to squeeze through capillaries in order to deliver O_2 and pick up CO_2. Think of a RBCs as a partly inflated beach ball. When a beach ball is completely inflated, you can't really move it around or put it in the car safely. If you take some of the air out you can bend it, twist it, and stow it away wherever! This is going to be an important characteristic because, in order to get through the capillaries, RBCs have to squeeze through in a single file (one at a time). As they squeeze through, they actually release ATP and other messages that tell the capillaries to dilate and open up a little to allow them to pass. If RBCs are too big or are not flexible enough, the task of going through the capillaries becomes extremely difficult.

Clinical Correlate

Hereditary Spherocytosis

There are various hereditary diseases that can affect RBC function. Among these is Hereditary Spherocytosis or HS. In HS, mutations in ankyrin, β spectrin, band 3 protein, α spectrin, and protein 4.2 lead to a decrease in the size of the membrane of the RBC. Remember our beach ball example? Well, RBCs with a decreased membrane surface area behave like beach balls that are completely inflated! As you can imagine, they are not as flexible as normal RBCs. This leads to hemolysis (medical term for the breaking up of RBCs) and a decreased amount of RBCs in the blood, which is called anemia. Under the microscope, these RBCs look like spheres, and this is where the disease gets its name!

Another amazing feature of RBCs is that they do *not* have mitochondria! This means that even though they carry oxygen, they don't need to consume it for energy. They instead produce energy from glucose via glycolysis. You wouldn't want your postal driver to be opening and using the packages you purchased, would you? In fact, mature RBCs contain no nucleus, and no organelles at all! This allows them to carry a lot more hemoglobin and use very little energy. This also means they can't be targeted by viruses, which by definition need to use a cell's processes to multiply and spread. Mature RBCs don't divide because of the aforementioned lack of a nucleus/organelles. They are created with nuclei inside the bone marrow so they can create the proteins and such to form a fully functional cell, but they lose the nuclei and the organelles as they mature. RBCs offer their services for a limited time, usually around 100 to 120 days, but in certain disease states live much shorter life spans, such as in hereditary spherocytosis (see the Clinical Correlate, *Hereditary Spherocytosis*). In fact, if you see lots of RBCs with nuclei, this can often be a sign of rapidly increased bone marrow red cell production, which is seen in states where the RBCs are breaking down faster than they can be made by the bone marrow. So, when we account for all of these features that make the RBCs oxygen carriers, you can see that nature outdoes the post office. Nature's trucks are more like aerodynamic disposable tanks that are flexible enough to squeeze through tight spaces!

Clinical Correlate

Hemoglobin and Hematocrit

In the hospital, when you want to analyze the contents of a person's blood, you order something called a complete blood count (CBC). The patient's blood will be drawn and spun down so that the heaviest parts (RBCs) accumulate in the bottom. The plasma, which is less dense, floats to the top and has a cloudy/yellow/straw-colored appearance (Figure 5.1A). The hematocrit is simply the percentage of stuff at the bottom. Since approximately 99% of that stuff are RBCs, the hematocrit serves as a measure of RBC content within the total blood. And since hemoglobin is very abundant in the RBCs, hematocrit is an indirect measure of hemoglobin content. The CBC will also report the amount of hemoglobin present in the sample in grams per deciliter or g/dL. The normal amount of hemoglobin varies between 12 and 15 g/dL, depending on sex and age.

Clinical Correlate

Anemia

Anemia is simply a low RBC count (and thus low hemoglobin content). Anemia can be acute (which means that something recently happened that decreased the amount of RBCs in that patient) or chronic (there's a long standing problem that is altering RBC production or lifespan). Acute anemia is generally due to hemolysis (breakdown of RBCs in the blood vessels or in the spleen) or to bleeding. (Keep in mind that in order for there to be anemia after bleeding, you need to recover some fluid *without* increasing the RBC content). Think of it like this: If you pour out half a bottle of soda and then measure the amount of sugar in the soda that's left in the bottle it will be the same regardless of the amount of soda. However, if you refill the half-filled soda bottle with water and then measure the amount of sugar, it will be decreased. The same thing happens when someone bleeds. You need the body to recover some fluid to dilute the RBCs that are floating around! Chronic anemia can result from multiple causes including problems with manufacturing RBCs (e.g., aplastic anemia), problems with manufacturing the hemoglobin within them (e.g., sickle cell anemia, thalassemias), RBC loss (e.g., chronic blood loss such as a gastrointestinal bleed), problems with RBC shape (e.g., hereditary spherocytosis), and problems with RBC glycolysis (e.g., G6PD, pyruvate kinase deficiency), among others. Problems with enzymes that are involved in glycolysis affect RBCs much more than other cells because RBCs don't have mitochondria and therefore exclusively rely on glycolysis for energy.

Oh MARVELOUS HEMOGLOBIN!

As we said earlier, due to the low solubility of O_2 in plasma, blood plasma is *not* enough to provide oxygen to all the cells in the body. But our delivery vehicles, RBCs, have a secret weapon: hemoglobin! Hemoglobin is a large iron-containing protein that can transport both O_2 and CO_2, independent of their solubility in plasma. This means that the more hemoglobin we have, the larger our capacity to transport O_2 and CO_2 in the blood. (Although we could go into great length about how it is manufactured and utilized, it's really not germane to understanding O_2 delivery per se. So we'll try and keep this short and sweet.)

When first made in the bone marrow, the RBCs still have their machinery; that is, they have a nucleus and organelles, and they consume oxygen to efficiently make ATP all with the single purpose of making massive amounts of hemoglobin ... so much so that by the time they're mature, 96% of their dry weight is made up of hemoglobin! As they are released into the blood by the bone marrow they get rid of all their internal machinery and dedicate all ~ 120 days of their life to delivering oxygen.

Hemoglobin is a rather complex molecule made up of four ring-shaped units of something called "heme." Each unit of heme has iron in the Fe^{2+} (ferrous) state, which acts like a sort of oxygen magnet scooping up oxygen. When oxygen gets close, it for forms a temporary bond with the iron in the hemoglobin. This allows hemoglobin to snatch up oxygen molecules and hold onto them, thus removing O_2 from solution and driving further O_2 into solution via diffusion. Since each hemoglobin molecule is made from four different chains, and each chain can bind one molecule of O_2, each hemoglobin molecule can bind four molecules of O_2.

Hemoglobin's interaction with O_2 hinges on the concept of affinity. Affinity is the property by which different chemical species bind to form chemical compounds. In other words, how easy it is for two dissimilar things to bind! Hemoglobin's affinity for O_2 is variable, meaning some things make hemoglobin want to bind more readily to O_2 (increase affinity), while others make hemoglobin want to let go of O_2 (decrease affinity). This is a critical component of transporting O_2 to and from tissues. Additionally, something that's particularly fascinating about hemoglobin's interaction with O_2 is that it displays something called *cooperativity*. When hemoglobin contains no O_2, it's a little bit harder for O_2 to bind to any one of the four heme subunits. However, as O_2 begins to bind to the iron, there's actually a structural change in the hemoglobin such that each O_2 molecule that binds to a heme subunit makes subsequent binding of more O_2 easier and easier. Thus hemoglobin that has one O_2 molecule bound has a higher affinity for O_2 than hemoglobin with no O_2. Hemoglobin with two O_2 molecules bound has a higher affinity that that with one, and so on. This co-operative change in affinity is so great that a hemoglobin molecules with three heme subunits bound to oxygen has an affinity 300 times greater than the hemoglobin that has none bound. This phenomenon is graphed out in the O_2–hemoglobin dissociation curve (Figure 5.3).

Figure 5.3 is a complex figure so we'll walk you through it. The X-axis represents the partial pressure of O_2, and we have two Y-axes,

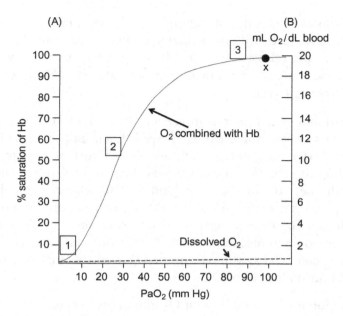

Figure 5.3 The O₂ hemoglobin dissociation curve.

one on the left **(A)**, which represents the % saturation of Hb with O_2 (which means of all the available sites for O_2 binding, what % are actually occupied?) and one on the right **(B)**, which is the actual amount of O_2 in mL per deciliter (dL, 100 mL) of blood. You'll notice that there are two lines drawn in the graph. Take a look at the dotted line in the bottom labeled "dissolved O_2." Remember how at the beginning of this chapter we mentioned how almost no O_2 travels in the blood as dissolved O_2? Well, what this line represents is the amount of O_2 that there would be in plasma if there was no hemoglobin and we relied only on the dissolved O_2 to get the job done. If you take a look at the total amount of dissolved O_2 that is being transported you'll realize that it is close to nothing!

Let's put some actual numbers behind this assertion. The amount of blood that the heart pumps out in 1 minute is known as the cardiac output. It is approximately 5 L, therefore we can say that in steady state conditions cardiac output is 5 L/min. In other words, 5 L is the amount of blood that circulates around the body in 1 minute. So, looking at our right Y-axis **(B)**, assuming our P_AO_2 is 100 mmHg, and a solubility of O_2 of 0.003 mL/mmHg, then the amount of dissolved O_2 dissolved in plasma would be 0.3 mL/dL of plasma (100 mmHg $O_2 \times 0.003$ mL/mmHg $= 0.3$ mL/dL). Since there are 10 dL in 1 L, then there are 50 dL in 5 L. This means that if our cardiac output is 5 L/min, and we have 0.3 mL/dL of O_2, then in the entire cardiac output there are 15 mL of O_2 (0.3 mL/dL $\times$ 50 dL = 15 mL of O_2). Remember from last chapter when we said that the consumption of O_2 for the body is approximately 250 mL of O_2 per minute? Well, 15 mL doesn't even come close! Clearly, RBCs and hemoglobin are absolutely essential for oxygen delivery in the human body.

Take a look at the line labeled "O_2 combined with Hb" in Figure 5.3. This is the actual O_2–hemoglobin dissociation curve, and it represents the amount of O_2 in the blood that is bound with hemoglobin at different pressures of O_2. The axes are the same for the amount of dissolved O_2 in blood. However, unlike the dissolved O_2 line, which is a straight line, the O_2–hemoglobin dissociation curve is not straight at all. In fact, it's a sigmoid curve. What does this mean? Well, remember when we talked about cooperativity in the previous section? This is where we can see it in all its glory. We said that cooperativity is the phenomenon through which the binding of one O_2 molecule increases the likelihood that more O_2 will bind to the hemoglobin. Think of it as a party—if you were looking for something to do on a Friday night, would you like to go to a party where there is only one other person? Not really... But what if we invited you to a party that has 20 to 30 people? This sounds a little more appealing, no? This is cooperativity. The same applies for the O_2–hemoglobin relationship. The more O_2 that's already bound to hemoglobin, the more O_2 that will bind to it.

Now, if we take a look at Figure 5.3 you'll see that the initial part of our curve is a little flat, then it gets steep, and then it goes flat again. What does this mean? Well, it's basically the plotting out of the cooperativity phenomenon. Initially, when there is no O_2 bound to hemoglobin, it's difficult to bind the first molecule of O_2. Therefore the pressure of O_2 has to increase a lot in order to start binding O_2 to hemoglobin (flat part labeled 1). Once the first O_2 is bound, then it becomes progressively easier to bind more O_2. This means that the pressure of O_2 only has to increase slightly to increase the amount of O_2 bound to hemoglobin (steep part labeled 2). Once the pressure of O_2 starts exceeding approximately 75 mmHg the curve flattens out again (flat part labeled 3). This happens because now, hemoglobin is already bound to a large amount of O_2, so even large changes in O_2 pressure have a relatively small effect on the amount bound to hemoglobin. (Take into account that there are approximately 250 million hemoglobin molecules per RBC. So the O_2–hemoglobin dissociation curve is an average of all hemoglobin molecules. With 250 hemoglobin molecules per RBC, each molecule binding four oxygen atoms, it means that there are roughly one *billion* oxygen molecules carried per individual RBC!)

Key

Ninety-nine percent of the oxygen in blood is bound to hemoglobin, and only 1% travels freely dissolved in plasma.

O_2 CONTENT AND O_2 DELIVERY

This graph is nice, but is there a formula that we can use to actually quantify how much O_2 a patient's blood is actually carrying? We're so glad you asked. Yes, there is! It's called the O_2 content equation and it's calculated as follows:

$$O_2 \text{ blood content in mL/dL} = \text{Hb(g/dL)} \times 1.34 \text{ mL/g} \times \text{Saturation of Hb}$$

where:

Hb = Hemoglobin in grams per deciliter (dL)
1.34 mL/g = The maximum amount of O_2 that 1 gram of hemoglobin can carry when saturated at 100%
Saturation of hemoglobin = The percentage of O_2 carrying sites that are currently carrying O_2 (this value is expressed as a decimal)

If you think about it, the first two terms actually give us the maximum amount of O_2 that our patient's blood can carry! When we multiply times the saturation, what we're doing is calculating how much the blood is actually carrying.

So, if we assume a concentration of Hb of 15 g/dL and a saturation of 99%, how much O_2 is the blood carrying?

$$O_2 \text{ blood content in mL/dL} = 15 \text{ g/dL} \times 1.34 \text{ mL/g} \times 0.99$$
$$O_2 \text{ blood content in mL/dL} = 20.1 \text{ mL/dL} \times 0.99$$
$$O_2 \text{ blood content in mL/dL} = 19.9 \text{ mL/dL}$$

Does this correspond with what we see in Figure 5.3? In fact it does! If you take a look, point X roughly approximates all these values. So the O_2 blood content formula allows us to quantify how much O_2 is in a given blood sample where we know the hemoglobin and at a specific saturation. (Keep this equation in mind, because we will come back to it in a little bit.)

Before we get crazy with numbers, let's note a couple of things that are really key to understanding the O_2–hemoglobin dissociation curve.

- Changes in O_2 pressure that are quantitatively the same (e.g., a decrease in 20 mmHg of O_2) can have a completely different effect on the amount of O_2 being carried by the hemoglobin. A decrease from 100 mmHg to 80 mmHg decreases the saturation only slightly (e.g., from 99% to 92%, a 7% drop), whereas a decrease in pressure from 60 mmHg to 40 mmHg, although quantitatively the same (e.g., 20 mmHg) would decrease saturation from around 88% to 72%, a 16% drop, more than twice as much as before! This is why we can be relatively comfortable with patients that have O_2 saturations above 92%, because this means that we're functioning on that flat part of the O_2–hemoglobin dissociation curve. But as soon as the saturation starts to drop below that, red flags should go up immediately! Think about it: Once O_2 saturation starts to drop and we move to the steep part of the curve, O_2 saturation can drop and drop fast.
- The O_2–hemoglobin dissociation curve is *independent* of the *amount* of hemoglobin. This is important to consider, because a normal patient with 15 g/dL of hemoglobin can have the same saturation of hemoglobin (e.g., 99%) as someone with anemia and a hemoglobin concentration of 7 g/dL (saturation can also be 99%). Think about this with regard to the O_2 content formula; an 8 g/dL drop in hemoglobin will result in having around 20 mL/dL of O_2 to 9 mL/dL of O_2! So, remember, a high saturation does not necessarily mean adequate O_2 content.

Key

The saturation of hemoglobin is *independent* of the amount of hemoglobin that is circulating in the blood. A patient with anemia can also saturate at 99% and still have poor O_2 carrying capacity.

Now we know how to calculate the content of O_2 in the blood. However, out of the O_2 that is being carried by the blood how much is actually getting to the tissues? Well, if we already know how many mLs of O_2 the blood has in 1 dL (100 mL), then all we're missing is how much blood is circulating around the body. This is where cardiac output comes in! As we said previously, cardiac output is the amount of blood that the passes through the heart in 1 minute, and it is approximately 5 L. So, now that we know this, let's calculate the delivery of O_2. This can be done with the O_2 Delivery equation (DO_2), which is basically a mash together of O_2 content and the cardiac output:

$$DO_2 = CO \times O_2 \text{ Content}$$

or

$$DO_2 =$$
$$CO \times O_2 \text{ blood content in mL/dL} = Hb(g/dL) \times 1.34 \text{ mL/g} \times \text{Saturation of Hb}$$

where:

CO = Cardiac output
Hb = Hemoglobin in grams per deciliter (dL)
1.34 mL/g = The maximum amount of O_2 that 1 gram of hemoglobin can carry when saturated at 100%
Saturation of hemoglobin = The percentage of O_2 carrying sites that are currently carrying O_2 (this value is expressed as a decimal)

All we're doing with this formula is multiplying the quantity of O_2 in 100 mL of blood (as we said, 20 mL O_2/dL of blood) times the quantity of blood that circulates in the body (5 L per minute)! That's it. The key when actually calculating this formula is getting the units right because cardiac output is in L/min and O_2 content is in mL/dL, so you need to convert either the cardiac output to dL or the O_2 content to L. It is our preference to convert the cardiac output to dL, so considering there are 10 dL in 1 L, then you multiply the cardiac output times 10. So, how does the formula look when we plug in the numbers assuming 5 Liters of cardiac output (i.e., 50 dL), 15 g/dL of hemoglobin, and a saturation of 99%?

$$DO_2 = 50 \text{ dL/min} \times (15 \text{ g/dL} \times 1.34 \text{ mL/g} \times 0.99)$$

Calculating the content of O_2 first:

$$DO_2 = 50 \text{dL/min} \times 20 \text{ mL } O_2/\text{dL}$$

so:

$$DO_2 = 1000 \text{ mL } O_2/\text{min}$$

Therefore, in this case, the delivery of O_2 to the tissues is approximately 1000 mL O_2/min, which, considering our baseline consumption of 250 mL O_2/min, should be more than enough! An important concept to understand is that the DO_2 formula defines the three things that can be done to increase the delivery of O_2 to tissues. We can:

• Increase the cardiac output
• Increase the amount of hemoglobin in the blood
• Increase the saturation

That's it, there's nothing else that we can do in order to increase the amount of O_2 that is delivered to tissues! Therefore learning how and when we should alter any of these variables—cardiac output, hemoglobin content, and saturation—is of paramount importance.

Key

Only three things can be done to increase the delivery of O_2: (1) increase cardiac output, (2) increase the amount of hemoglobin, and (3) increase the saturation of hemoglobin with O_2.

Clinical Correlate

Shock

Shock is defined as generalized tissue hypoperfusion—in other words, tissues not getting enough oxygen! There are several kinds of shock (e.g., hypovolemic, neurogenic, cardiogenic, and septic), but in all of these conditions the delivery of O_2 to the tissues is compromised. In hypovolemic, neurogenic, and cardiogenic shock there's a problem with the actual moving of blood from the heart and lungs to the tissues. This means that cardiac output is low, and if cardiac output is low the delivery of O_2 is also low! In hypovolemic shock there isn't enough blood in the system to delivery adequate amounts of O_2. In neurogenic shock, the blood is there, but it's pooled in the veins and can't make it back to the heart and lungs. In cardiogenic shock, since the heart isn't functioning properly all the blood is pooling behind the heart (this is why patients can present with pulmonary edema!). Septic shock is a different beast all by itself because the problem lies in the systemic capillaries, where O_2 exchange is

> impaired, so even though there might be a good delivery, the availability of O_2 for tissue uptake is reduced. The bottom line is that in all of these conditions, the amount of O_2 that is getting to the tissues is low!

So, this is why we need hemoglobin. It increases the ability of blood to carry O_2 throughout the body. OK, that makes sense, but then, how does the hemoglobin know when to hold onto O_2 and when to release it? It would be of no use to have hemoglobin that is supersaturated with O_2 floating around the body but unable to deliver the O_2 to where it's actually needed. Hence various factors can alter hemoglobin's affinity for O_2. In fact these mechanisms are so efficient that affinity for O_2 increases when hemoglobin is passing through the pulmonary circulation and decreases in the systemic capillaries. How exactly does that happen?

Clinical Correlate

Oxihemoglobin and Carboxihemoglobin

When hemoglobin is bound to O_2 it's known as oxihemoglobin. This is the type of hemoglobin that we want because it can load O_2 in the lungs and offload O_2 in the peripheral tissues. Carboxihemoglobin on the other hand is the term given when carbon monoxide binds to hemoglobin. Unlike O_2, carbon monoxide is toxic. It binds the iron in hemoglobin at the same site the O_2 does (hence the name carboxihemoglobin) but unlike O_2, the affinity of iron for carbon monoxide is about 200 times that of O_2. Translated into the clinical setting this means that once hemoglobin is bound to carbon monoxide its ability to bind to O_2 and deliver it to tissues is compromised. Carbon monoxide is one of the many products of combustion. Therefore people who are at increased risk for carbon monoxide poisoning are those who spent large amounts of time in confined spaces where combustion is occurring (e.g., victims of house and building fires, people who barbeque indoors, etc.). The treatment for carbon monoxide poisoning is increasing the FiO_2 in order to displace the carbon monoxide from the hemoglobin.

DYNAMICS OF O_2–HEMOGLOBIN DISSOCIATION CURVE: HOW DOES IT KNOW WHERE TO DELIVER ITS CARGO?

As we saw in the previous section, hemoglobin is the main carrier of O_2 in the body. As is intuitive, when the RBCs traverse the pulmonary capillaries they load up with O_2 and then they head straight toward the heart, out through the aorta, through progressively smaller arteries, and then into the capillaries. But the O_2 is still bound to the hemoglobin, so how the heck are we going to get the O_2 from the hemoglobin into the tissues? How about if we could change the affinity of the hemoglobin so that affinity for O_2 increases in the lungs (which improves binding of O_2) and then magically decreases in the peripheral tissues (which would the release O_2)? Well, Mother Nature got it right again, because this is exactly what happens!

One of the coolest aspects of hemoglobin is that the affinity the iron ions have for oxygen changes depending on the environment. In traditional textbooks, you'll find a list of things that increase the affinity of hemoglobin for O_2 (i.e., left shift the curve) and a list of things that decrease the affinity of hemoglobin for O_2 (i.e., right shift the curve). We're going to take a slightly different approach. (We will include the list, don't worry!) Think about it like this: When do you need hemoglobin's affinity to be the highest? In the lungs, of course—this will favor uptake of the O_2 from the alveolus and into the RBC. OK, when do you need hemoglobin's affinity to be the lowest? In the tissues where the O_2 needs to be delivered, because it is there that it is being consumed! And why is O_2 being consumed? Because cells are creating energy; that is, there are ongoing metabolic requirements. This means that cells will consume O_2, but as they do they will dump out CO_2 (which increases the pressure of CO_2 in the interstitial space), they will dump out H^+ (which decreases the pH), and the temperature will increase! I think you can see where we're going with this. If you increase metabolism, you increase O_2 consumption, and if you increase O_2 consumption, you increase the byproducts of metabolism, namely CO_2, H^+, and temperature! So we could say that an increase in the CO_2, H^+, and temperature roughly equate to an increase in O_2 consumption. Any increase in consumption has to be met with an increase in delivery.

Therefore the metabolic byproducts that signal an increase in tissue metabolic activity and in O_2 consumption are CO_2, H^+, and temperature. As these compounds increase, the affinity of hemoglobin for O_2 decreases! This means that the more CO_2 and H^+, and the higher the temperature, the easier it is for hemoglobin to offload O_2 and deliver it to the tissues where it is actively being consumed. In terms of our O_2–hemoglobin dissociation curve, this is called a right shift. Conversely, when the byproducts of metabolism are low, hemoglobin's affinity for O_2 increases; that is, it binds to O_2 more readily—this is called a left shift. Why does this happen? Well, where would you want hemoglobin's affinity for O_2 to be the highest? In the lungs, where hemoglobin has to bind to O_2. And, where would you want hemoglobin's affinity for O_2 to be the lowest? In the peripheral tissues where hemoglobin needs to offload O_2. Accordingly, in the lungs the pressure of CO_2 decreases because it's being exchanged with the atmosphere; the concentration of H^+ decreases as well because of the decrease in CO_2, and the temperature also decreases because some heat escapes into the alveolar air.

Take a look at Figure 5.4, which expresses all of this in graphical format. In Figure 5.4, **Curve B** is our reference O_2–hemoglobin dissociation curve. As we mentioned, if we increase O_2 consumption and therefore increase the byproducts of metabolism, the affinity of hemoglobin for O_2 decreases; that is., there is a right shift (**Curve C**). Alternatively if the byproducts of metabolism decrease, the affinity of hemoglobin for O_2 increases (**Curve A**). An easy way to gauge differences in affinity is by comparing the saturation percentage of all three

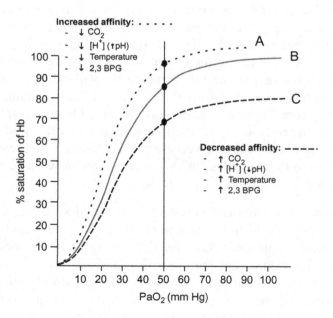

Figure 5.4 The affinity of hemoglobin for O_2 can be modified by several factors including CO_2, $[H^+]$, temperature, and 2,3 BPG.

curves at the same pressure of O_2. The black line placed at a pressure of 50 mmHg in Figure 5.4 does just that. You can clearly see that at the same pressure of O_2, the approximate saturations of each curve decrease as affinity decreases. One additional factor that modifies the affinity of hemoglobin to O_2 is 2,3 biphosphoglycerate (2,3 BPG). 2,3 BPG is an end product of metabolism in the RBC, which means that the more metabolically active an RBC is, the higher the concentration of 2,3 BPG is going to be. Similar to what we've been discussing thus far, increases in 2,3 BPG decrease the affinity of hemoglobin for O_2. Conversely decreases in 2,3 BPG increase the affinity of hemoglobin for O_2.

There are a couple of interesting additions to the previous concepts that are worth mentioning. If you noticed, at *no* point in the previous discussion did we use the term pH; instead we used the $[H^+]$ as a surrogate. So keep in mind that $[H^+]$ and pH vary inversely. As $[H^+]$ increases, pH decreases, and vice versa. Now, why exactly do H^+ and CO_2 change the affinity of hemoglobin for O_2? As either H^+ or CO_2 bind to hemoglobin there will be a change in the shape of the hemoglobin that decreases the affinity of hemoglobin for O_2. The particular effect that changes in $[H^+]$ have on hemoglobin affinity for O_2 is called the Bohr effect, after Danish physiologist Christian Bohr (father of physicist Niels Bohr). CO_2 has a similar effect on hemoglobin, in that higher pressures of CO_2 decrease affinity for O_2 and consequently decrease the binding of O_2 to hemoglobin. This is called the Haldane effect, after Scottish chemist John Scott Haldane.

Key

Increases in CO_2, $[H^+]$, 2,3 BPG, or temperature decreases the affinity of hemoglobin for O_2.

Key

The Bohr effect describes how pH modifies hemoglobin's affinity for O_2; the Haldane effect describes how CO_2 modifies hemoglobin's affinity for O_2.

This is all very nice, but what exactly is going on at the level of the cell? Take a look at Figure 5.5. This figure shows that in the

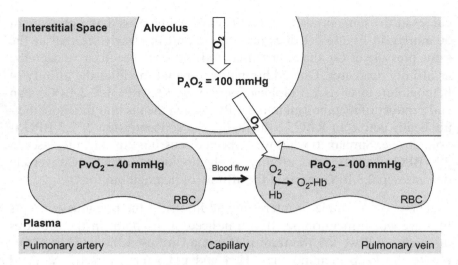

Figure 5.5 In the pulmonary capillaries the increased pressure of O_2 in the alveoli relative to the RBCs favor the diffusion O_2 into the RBCs where O_2 binds hemoglobin and forms oxyhemoglobin.

deoxygenated blood arriving from the venous circulation, the pressure of O_2 is around 40 mmHg. Once it reaches the alveoli where the pressure is 100 mmHg, O_2 diffuses from the alveolus to the RBC and binds hemoglobin to form oxyhemoglobin (O_2-Hb). This increases the pressure of O_2 in the oxygenated blood to the same level as in the alveoli. (This is why a low alveolar pressure of O_2 prevents adequate oxygenation of the peripheral tissues!) Once the blood leaves the lungs and gets to the heart, there's mixing of blood that was fully oxygenated (PaO_2 of 100 mHg) and blood that wasn't fully oxygenated, which decreases the PaO_2 to 95 mmHg. The blood that wasn't fully oxygenated is a combination of venous blood from the bronchial circulation and differences in the ventilation/perfusion ratios in the lung as we'll see later on. Even though the PaO_2 is around 95 mmHg saturation will still be close to 99% (remember cooperativity!). So the PaO_2 that arrives to the peripheral tissues carried by arterial blood is close to 95 mmHg (Figure 5.6).

When the O_2 laden RBCs reach the peripheral tissues, they will start exchanging O_2 with the interstitial fluid where the pressure of O_2 is close to 40 mmHg. The difference between the PaO_2 and the interstitial fluid is 55 mmHg (95 mmHg − 40 mmHg). This amounts to a 55 mmHg gradient that favors diffusion of O_2 from the RBC to the interstitial fluid! (Inside the cells the pressure of O_2 is around 20 mmHg, so there's still a

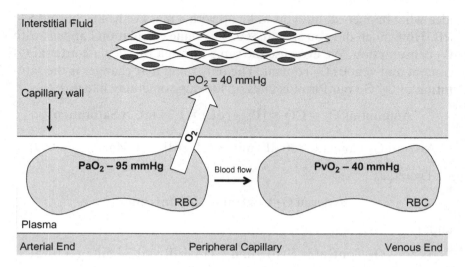

Figure 5.6 In the peripheral capillaries, the high pressure of O_2 in the blood relative to the interstitial space favors diffusion of O_2 from the RBC to the interstitial fluid.

pretty big gradient to help get O_2 from the interstitial space into the cells!) This means that after exchanging O_2 with the interstitial fluid, the pressure of O_2 in the venous blood, or PvO_2, will be the same as that in the interstitial fluid, 40 mmHg. Venous blood with a PvO_2 of 40 mmHg then travels back to the heart and lungs to get reoxygenated and repeats the cycle. (It is important to keep in mind that if O_2 consumption increases suddenly, cells will start to consume the O_2 in the interstitial fluid, which will decrease the pressure of O_2 in the interstitial space. If the pressure of O_2 in the interstitial space decreases, then the gradient between arterial blood and interstitial fluid increases, favoring a larger O_2 extraction by the tissues.)

So how do we know how much O_2 is being consumed by the peripheral tissues? Considering that with our O_2 delivery equation we calculated how much was arriving it shouldn't be too complex to calculate how much is being taken up by the tissues, should it? It actually isn't! And as it turns out it's relatively easy. We use this formula:

$$O_2 \text{ consumption} = \text{Arterial } O_2 \text{ content} - \text{Venous } O_2 \text{ content}$$

In simple terms, it subtracts the amount of O_2 that returns to the lungs (Venous O_2 content) from what originally left (Arterial O_2 content). Think about it like this: You decide to go out and party like crazy one night and you have $100 on you. The next morning, you have no idea what happened, but you wake up with a massive headache and $40 left. How much did you spend? $60. The same calculation happens with O_2 consumption. We even use the same formula to calculate arterial O_2 content and venous O_2 content. The only thing that changes is the saturation of O_2. So our formula ends up looking something like this:

$$\text{Amount of } O_2 = CO \times Hb \, (g/dL) \times 1.34 \, mL \times \text{Saturation}$$

$$\text{Arterial } O_2 \text{ Content} = 50 \, dL/min \times (15 \, g/dL \times 1.34 \, mL/g \times 0.99)$$

Therefore:

$$\text{Arterial } O_2 \text{ Content} = 1000 \, mL/min$$

and:

$$\text{Venous } O_2 \text{ Content} = 50 \, dL/min \times (15 \, g/dL \times 1.34 \, mL/g \times 0.75)$$

(mixed venous blood saturation is approximately 75%).

Therefore:

$$\text{Venous } O_2 \text{ content} = 750 \, mL/min$$

and:

$$O_2 \text{ consumption} = 1000 \, mL/min - 750 \, mL/min$$

So:

$$O_2 \text{ consumption} = 250 \, mL/min$$

This value of 250 mL/min is what we saw in Chapter 4 as the standard O_2 consumption in 1 minute! So now you know where we got this number from.

One caveat to this is where you sample the venous blood from (and it's a big caveat so be careful). Mixed venous blood is the term used for venous blood from the entire body that has been mixed so that there's an equal contribution from all parts of the body. If you were to sample blood that is coming from a body segment that is really metabolically active and therefore extracting more O_2, the saturation will be lower. Conversely if you sample venous blood from a segment that is not consuming that much O_2, the venous saturation will be higher. In both cases the calculation of O_2 consumption will be off, so be careful where the venous blood is coming from.

TRANSPORT OF CO_2 IN THE BLOOD

CO_2 transport in blood is a little different than O_2 transport. Unlike O_2, which travels almost 99% bound to hemoglobin in the blood, the majority of the CO_2, almost 70%, travels as bicarbonate (HCO_3^-), approximately 25% travels bound to hemoglobin, and 5% travels dissolved in plasma. So, how on earth is CO_2 being mostly transported as HCO_3^-? Let's take a look at a chemical reaction, with which you will become very familiar. It's the reversible binding of H_2O and CO_2, which yields carbonic acid (H_2CO_3), which in turn dissociates into H^+ and HCO_3^-.

$$H_2O + CO_2 \leftrightarrow H_2CO_3 \leftrightarrow H^+ + HCO_3^-$$

This reaction is powered by an enzyme called carbonic anhydrase. Carbonic anhydrase accelerates the binding of CO_2 and H_2O to form H_2CO_3. We will see this reaction and its effects on pH in a little more detail in Chapter 7. However for now, let's briefly touch on what determines the direction in which the reaction takes place; that is, does it produce H_2O and CO_2 or H^+ and HCO_3^-? Since all the steps in this reaction are reversible, the concentration of the substrate is what determines the product. Wait, what? Think about it like this: If there's an increase in the amount of CO_2 it will push the reaction toward the formation of H^+ and HCO_3^-. If, however, there's an increase in amount of H^+ it will drive the reaction in the other direction, toward the formation of H_2O and CO_2. That said, take a look at Figure 5.7; in it you can see that as RBCs traverse the peripheral capillaries from the arterial end to the venous end, as cells consume O_2 they produce CO_2 and the pressure of CO_2 increases from 40 mmHg to 45 mmHg. This is because CO_2 diffuses readily across the capillary wall, and since the pressure of CO_2 in the interstitial fluid is 5 mmHg higher than that of the RBCs (45 mmHg vs 40 mmHg), the CO_2 diffuses into the RBC. (We had previously mentioned that the gradient for O_2 was close to 50 mmHg, and the gradient for CO_2 is about 10 times less. This is because CO_2 diffuses a lot faster than O_2!)

Inside the RBC, things get even more interesting (Figure 5.8). As the pressure of CO_2 increases in the interstitial fluid, CO_2 will diffuse into the RBC. Once in the RBC there are two pathways that CO_2 can follow: (1) CO_2 can directly bind hemoglobin and form CO_2-Hb (carbaminohemoglobin) or (2) it can react with H_2O to create H_2CO_3, which will then dissociate into H^+ and HCO_3^-. The H^+ ion will bind

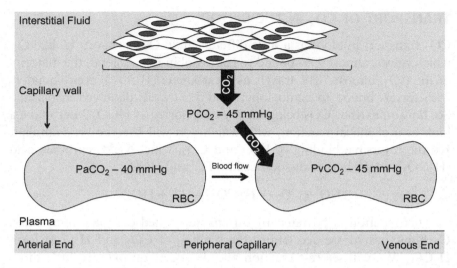

Figure 5.7 CO$_2$ is produced in the peripheral tissues, which increases the pressure of CO$_2$ in the interstitial fluid. CO$_2$ will then diffuse from the interstitial fluid to the RBCs.

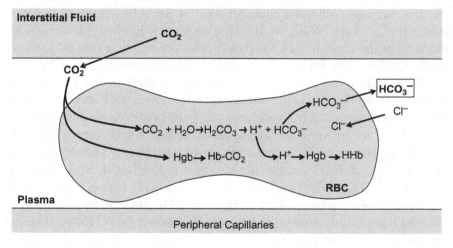

Figure 5.8 Once inside the RBC, CO$_2$ has two fates: (1) it can bind to hemoglobin to create carbamino-hemoglobin (Hb-CO$_2$) or it can react with water to create H$^+$ (which will bind to hemoglobin to form HHb and HCO$_3^-$), which will diffuse out of the RBC and will be exchanged with Cl$^-$.

to hemoglobin, creating HHb. The HCO$_3^-$ ion, however, will diffuse back into the plasma! (The HCO$_3^-$ that diffuses back into the plasma is exchanged with Cl$^-$ in order to maintain electroneutrality! This phenomenon is called the chloride shift.) As you can see, most of the CO$_2$ that is produced in the peripheral tissues and then diffuses into the RBC

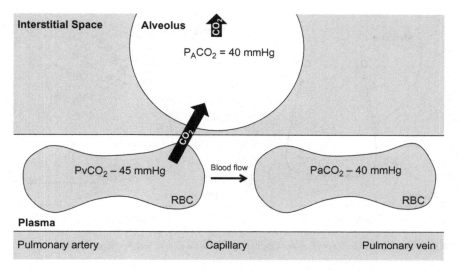

Figure 5.9 In the pulmonary capillaries, CO_2 diffuses from the blood to the alveolus where it is later breathed out.

turns into HCO_3^-, which is then shuttled back into the plasma once again! This takes place in the peripheral capillaries and venous blood. When the RBC gets to the lungs, the directionality of this process is reversed (Figure 5.9). Since the pressure of CO_2 in the alveoli (40 mmHg) is less than the pressure of CO_2 in the blood (45 mmHg), CO_2 diffuses from the blood to the alveoli, where it is later exhaled out to the atmosphere.

If we were to take all the previous concepts and mash them together into one simplified figure, it would look something like Figure 5.10. In it we can see how the RBC behaves simultaneously in both the pulmonary capillaries (A) and the peripheral capillaries (B). In the pulmonary capillaries the O_2 bind to hemoglobin to create oxyhemoglobin. This process dissociates the CO_2 from the hemoglobin, favoring diffusion toward the alveolus. In the peripheral capillaries, this process is reversed. The high pressures of CO_2 move into the RBC, displacing the O_2 from the hemoglobin and favoring the offloading of O_2 into the interstitial fluid for later uptake by the cells.

CLINICAL VIGNETTES

A 23-year-old man comes into the Emergency Department after being shot in the leg. He was reported to be bleeding profusely at the scene.

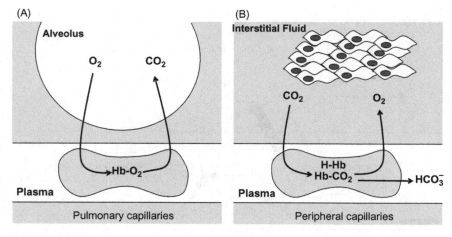

Figure 5.10 Integrated exchange of O_2 and CO_2 in the pulmonary capillaries and peripheral capillaries.

Paramedics administered 2 liters of 0.9% saline (normal saline) solution to him en route. Currently his vitals are as follows: HR 122, BP 90/60, RR 32, Temp 36°C, Saturation 99% with nasal cannula. You start infusing more saline solution, and draw some labs. The results come back:

Hemoglobin 10 g/dL (normal 13−15 g/dL)
Hematocrit 31% (normal 37−42)

1. Given that this patient lost a lot of blood and has had some IV fluid repletion, let's say that his cardiac output is a little lower than normal, say 4 L/min. What is the delivery of O_2 in this patient?
   A. Approx. 1000 mL/min
   B. Approx. 650 mL/min
   C. Approx. 530 mL/min
   D. Approx. 70 mL/min

Answer: C. In order to answer this question, we need to calculate the DO_2 according to our DO_2 formula:

$$DO_2 =$$
$$CO \times O_2 \text{ blood content in mL/dL} = Hb(g/dL) \times 1.34 \text{ mL/g} \times \text{Saturation of Hb}$$

where:

CO = Cardiac output
Hb = Hemoglobin in grams per deciliter (dL)

1.34 mL/g = The maximum amount of O_2 that 1 gram of hemoglobin can carry when saturated at 100%
Saturation of hemoglobin = The percentage of O_2 carrying sites that are currently carrying O_2 (this value is expressed as a decimal)

Therefore:

$$DO_2 = 40 \text{ dL/min} \times 10 \text{ g/dL} \times 1.34 \text{ mL/g} \times 0.99 \text{ saturation}$$
$$DO_2 = 530 \text{ mL/min}$$

2. Since the delivery of O_2 in this patient is decreased, the tissues could potentially start extracting O_2 at a higher rate from the O_2 that is delivered. A central line was placed in this patient and the mixed venous saturation of blood was measured at 50%. Using this information, calculate the amount of O_2 that is actually being extracted by the tissues:
   A. Approx. 260 mL/min
   B. Approx. 500 mL/min
   C. Approx. 150 mL/min
   D. Approx. 330 mL/min

Answer: A. In order to get to the amount of O_2 that is extracted, you must first calculate the content of O_2 in the venous blood using a saturation of 50%. Using the same DO_2 formula but switching the saturation from 0.99 to 0.50, this will be 268 mL/O_2/min. This number represents the amount of O_2 in the venous blood. To calculate the extraction rate, we need to subtract the venous content from the arterial content, so: 530 mL/min − 268 mL/min, which is approximately 260 mL/min.

CHAPTER 6

The Alveolar–Capillary Unit and V/Q Matching

So far in the book we've reviewed how air moves in and out of the lungs, and how the movement of air leads to changes in the pressures of O_2 and CO_2 in the alveolus. We then went ahead and discussed the enormously important role that blood has in transporting O_2 from the lungs to tissues and CO_2 from the tissues to the lungs. Let's add the next layer of complexity to this relationship!

THE ALVEOLAR–ARTERIAL DIFFERENCE: HOW GOOD IS THE LUNG AT EXCHANGING O_2 AND CO_2?

In Chapters 4 and 5 we learned the specific pressures of O_2 and CO_2 in the alveolus, the arterial blood, and the venous blood. But what good does it do us to know all these different P_A, Pa, and Pv's? (Remember: A = alveolar, a = arterial, v = venous.) In order to answer this, let's take a look at the functional unit of the lung: the alveolar capillary unit (Figure 6.1). As we've seen before, it is composed of the alveoli and the pulmonary capillaries that abut the alveolar wall. This is where the exchange of O_2 and CO_2 takes place. Look at Figure 6.1A to get a brief overview of the exchange process that's taking place at the level of the membrane. Blood that is coming in from the right side of the heart (blood from the venous side of the circulation) has a very low P_vO_2 40 mmHg (partial pressure of O_2 in the veins) and a high P_vCO_2 45 mmHg (partial pressure of CO_2 in the veins). This means that compared to the alveolar pressures, O_2 has a gradient of 60 mmHg from alveolus to the blood and CO_2 has a gradient of 5 mmHg from the blood to the alveolus. If our membrane works perfectly and the diffusion of O_2 and CO_2 happens without a problem, then the pressures of O_2 and CO_2 in the arterial blood will be identical to those in the alveoli. (Remember, diffusion happens until equilibrium has been reached; once the equilibrium has been reached—in this case the pressures of O_2 and CO_2 are equal on both sides of the membrane—diffusion will cease.) The membrane however, is not as simple as we would initially presume. In fact, it is made up of seven, yes, seven different layers, all of which

Back to Basics in Physiology. DOI: http://dx.doi.org/10.1016/B978-0-12-801768-5.00006-X

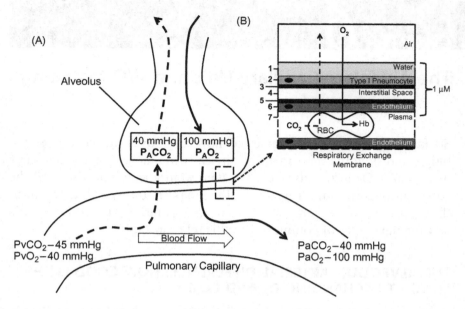

Figure 6.1 (A) The alveolar—capillary unit is the functional unit of the lung and it exchanges O_2 and CO_2 with the blood in the pulmonary capillaries. The gradient for diffusion of CO_2 from pulmonary capillaries to the alveoli is a 5 mmHg pressure difference (45 mmHg − 40 mmHg), and the gradient for O_2 diffusion from the alveoli to the pulmonary capillaries is 60 mmHg (100 mmHg − 40 mmHg). (B) A close-up look at the respiratory exchange membrane, which is approximately 1 µM thick and is composed of (1) Water lining the alveolar wall, (2) Type 1 Pneumocyte, (3) Pneumocyte basement membrane, (4) Interstitial Space, (5) Endothelial Basement membrane, (6) Capillary Endothelium, and (7) Plasma. (RBC − red blood cell).

are "sandwiched" together with an average approximate thickness of 1 µM (Figure 6.1B). What does this mean for exchange? Well, as we said earlier, if the membrane is working properly, there's no problem. However, anything that increases the thickness of any of the layers of the membrane will increase the distance that O_2 and CO_2 need to travel, thereby decreasing diffusion.

Key

Increasing the thickness of the membrane or decreasing the surface area available for exchange decreases diffusion.

As you can see, maintaining the P_AO_2 and the P_ACO_2 close to their target levels is extremely important to maintain a constant and adequate exchange of O_2 and CO_2 with the pulmonary capillaries. As we saw in Chapter 4, the alveolar gas equation

$$P_AO_2 = FiO_2(P_{ATM} - P_{H_2O}) - \frac{P_aCO_2}{RQ}$$

gives us a pretty good estimation of what was going on in the alveoli, but as we just saw, respiratory membrane function is also important. So, do you think there's something else that we can calculate in order to evaluate the function of our respiratory membrane? Yes there is! It's called the Alveolar–Arterial O_2 Difference, or $\Delta(A-a)$, and it's commonly referred to as the A–a gradient. Basically, in the simplest terms, this number is telling us the difference between the alveolar O_2 (P_AO_2) and the arterial O_2 (P_aO_2), and that's exactly how it's calculated:

$$\Delta(A - a) = P_AO_2 - P_aO_2$$

Key

The alveolar–arterial difference ($\Delta(A-a)$) is the difference between the P_AO_2 and the P_aO_2. It is a measure at how efficient the lungs are at exchanging O_2.

The normal value for the $\Delta(A-a)$ is actually a moving target, since it varies by patient, and there are many different reference ranges, especially in the elderly in whom the $\Delta(A-a)$ can be larger and be within normal limits. However, for the sake of simplicity, we'll establish that anything greater than a 10 mmHg difference warrants further workup as it could be potentially altered.

Now, *reader beware*! Despite it being referred to commonly as the A–a gradient, as we'll see in the following section the $\Delta(A-a)$ formula does *not* represent a *gradient*, it represents the *difference* in O_2 between *all* the alveolar air and *all* the blood going through the lungs. This is a crucial distinction. If you look at Figure 6.1A, why the $\Delta(A-a)$ is not a gradient is not really obvious. So what the heck is going on? Well, it's all a matter of perspective. The lungs are composed of millions upon millions of alveolar–capillary units, millions of little Figure 6.1 As put together, and the PaO_2 represents the average pressure of O_2 in all the arterial blood coming from all the pulmonary capillaries, not just a single unit. This means that the number we're actually measuring when we measure the PaO_2 is the average of *all* the alveolar capillary units put together, some that are super efficient and some that aren't as efficient. So in reality the $\Delta(A-a)$ has the O_2 of some alveoli that are awesome at their job (e.g., $PaO_2 = 100$ mmHg), and other alveoli that pretty much stink at it (e.g., $PaO_2 = 60$ mmHg). The blood coming from all these different alveoli gets mixed and the resultant is, for example, a PaO_2 of 80 mmHg. This doesn't mean that all the alveoli in the lungs have a gradient of 20 mmHg however (100 mmHg of P_AO_2 − 80 mmHg PaO_2); it means that we have some alveoli that are working and some that are not. It provides us the *whole* picture of how lungs are working.

VENTILATION/PERFUSION RELATIONSHIPS: MATCHING THE MOVEMENT OF O_2, CO_2, AND BLOOD

Why is all of this useful? This concept allows us to evaluate if the air we're bringing in is being matched with the blood that is circulating through the lungs. Wait, what? Think about it like this: If the goal of this entire system is to provide O_2 and remove CO_2, then blood and lungs have to function in unison in order to get the job done. The lungs move O_2 and CO_2 to and from the alveoli, while blood moves O_2 and CO_2 to and from the cells. Therefore we need to have a balance between V_A and blood flow in the lungs, which is called perfusion (Q). Tools like the Alveolar Gas Equation and the $\Delta(A-a)$ allow us to asses the relationship between Alveolar Ventilation (V_A) and perfusion (Q). When we talk about the V/Q relationship, we're talking about the balance between alveolar ventilation and blood flow to the lungs.

In very simple terms, the V/Q relationship establishes how efficient the lung is at matching the air being brought in to the blood that's passing through the lungs (Figure 6.2). An ideal relationship is 1:1, where the same amount of air and blood are in contact at the same time in the lungs to maintain O_2 and CO_2 within normal limits. So a V/Q ratio of 1 means that we have the same amounts of ventilation and perfusion. Anything other than a V/Q of 1 can be called a nonefficient use of resources. If ventilation is greater than perfusion (V > Q), there is too much air for the amount of blood that is in that alveolar capillary unit. If ventilation is less than perfusion (V < Q), there is too little air for the amount of blood that is in the alveolar capillary unit.

We can therefore theoretically divide the upright lung into three zones based on the previous information (Figure 6.2). In Zone 1, where

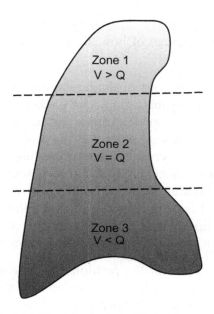

Figure 6.2 Ventilation (V), Perfusion (Q) relationships in the different lung zones.

$V > Q$, flow is intermittent (blood will flow only when the pressure of the pulmonary artery can overcome the airway resistance pressure). In Zone 1 all the blood that is going through is getting fully oxygenated, but it's quantitatively speaking only a small amount of blood compared to the blood that goes through Zones 2 and 3. The ideal V/Q ratio happens in Zone 2 where $V = Q$. This means that all the blood that goes through Zone 2 gets perfectly equilibrated because there is just the right amount of air for the blood that is going through. Zone 3, however, is a different story; thanks to the gravity that is pooling the blood in the bottom half of the lung there is too much blood compared to the amount of air available so the gases don't equilibrate as efficiently. Therefore in Zone 3, $Q > V$, which means that some blood is not going to offload all the CO_2 or bind all the O_2 it should. However, as we mentioned previously, the PaO_2 is a function of all three zones combined.

As we just saw, the blood from alveoli that are really good at exchanging, think Zone 1 or 2; mix their blood with alveoli that are not as efficient, think Zone 3. So, again, what's the relevance? Well, the alveolar gas equation and the alveolar—arterial difference allow us to evaluate the V/Q relationship of *all* the alveolar capillary units in the lung. This means that the $\Delta(A-a)$ will tell us how our GLOBAL V/Q is, since it's an average of how all the alveolar capillary units in the lung are working.

The most extreme versions of these relationships would be an area that is ventilated with *no* perfusion, $Q = 0$ (therefore $V/Q = \infty$). This would effectively function as dead space because air is moving in and out but no exchange is taking place. On the other end of the spectrum is an area that is *not* ventilated but is perfused, $V = 0$, (therefore $V/Q = 0$). This would effectively function as a shunt, in which venous blood mixes with arterial blood without exchanging O_2 and CO_2. In nondiseased states it is extremely rare to find either true shunt or true dead space ventilation. These extreme scenarios are generally present only when there is a significant physiologic derangement.

Now that we understand the relationship between the A−a difference and V/Q relationships, let's try and see why this is important. In people who are having trouble breathing and we need to try and narrow down where the problem is located, the A−a difference allows us to narrow down our options somewhat. Think about it like this: The A−a difference will allow us to evaluate how efficient our lungs are at exchanging O_2 and CO_2 with the blood (Figure 6.3). So, if our patient

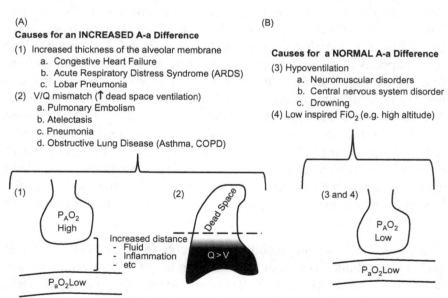

Figure 6.3 A−a differences in various diseased states.

is having difficulty breathing and the P_AO_2 is normal or high and the PaO_2 is low, this means that the A–a difference is increased; in other words, the O_2 in the alveoli is OK, but it's not making its way down to the blood, so we can very generally say that there's a problem with one of two things:

- There's an increase in the alveolar membrane thickness, which impairs diffusion. (This leads to a high P_aO_2 with a low P_aO_2, because diffusion is impaired. This is very rare in every day clinical practice.)
- There's a V/Q mismatch (this is a lot more common), with more areas with high blood flow and low ventilation, which basically means there's an increase in dead space ventilation. No matter how much air we put into dead space, there is no perfusion, so there can't be any exchange. In Figure 6.3, the diagram labeled (2) depicts this exact problem. This can happen in a number of diseases including pulmonary embolism, atelectasis (which is basically a collapsed segment of lung), pneumonia, and obstructive lung disease. Imagine that there's a pulmonary embolus that occludes perfusion to the top half of a lung. This means that no flow can take place in those segments of the pulmonary artery, effectively turning those segments into dead space ventilation. (All the air that goes into segments that are not ventilated is irrelevant when it comes to gas exchange.) Just like in traffic, when a street closes, all other streets get congested. The same thing happens to pulmonary blood flow, when a clot blocks off segments of the pulmonary artery. The blockage will divert the flow of blood to the rest of the lung, which means that perfusion (Q) is going to be too much for the ventilation (V). The final outcome is that there is not enough ventilation for the perfusion that is taking place, and although the P_AO_2 can be normal, the P_aO_2 is going to be low.

On the other hand, if the patient is having difficulty breathing and both the P_AO_2 and the PaO_2 are low (Figure 6.3B), this means that diffusion is happening appropriately, but if the P_AO_2 is low to start with, then the problem lies elsewhere—not enough O_2 is being brought into the lungs in the first place! The major causes for this are listed in Figure 6.3.

The relationship between ventilation and perfusion is incredibly more complex than what we've explained in this chapter. However, we believe that if you have a basic knowledge of the concepts outlined here it will be a lot easier to understand the pathophysiology of disease affecting lung function.

CLINICAL VIGNETTES

A 25-year-old male known heroin user is brought to the Emergency Department (ED) after being found unconscious at home, with an empty bottle of morphine tablets next to him. On arrival to the ED his PaO_2 is 40 and his $PaCO_2$ is 80.

1. What is his A—a difference?
    A. 10
    B. 30
    C. 50
    D. 80

Answer: A. In order to answer this question we must first calculate the P_AO_2 using the given information.

Considering:

$$P_AO_2 = FiO_2(P_{ATM} - P_{H_2O}) - \frac{P_aCO_2}{RQ}$$

Then:

$$0.21(760 - 47) - 80/0.8 = 50 \text{ mmHg for } P_AO_2$$

And in order to calculate the A—a difference we subtract

$$P_AO_2 - P_aO_2$$
$$50 \text{ mmHg} - 40 \text{ mmHg} = 10 \text{ mmHg}$$

Our A—a difference is 10 mmHg.

2. What is the likely cause of his presentation?
    A. Salicylate poisoning
    B. Barbiturate intoxication and central apnea
    C. Pneumonia
    D. This patient is healthy

Answer: B. Considering this patient's presentation, the most likely scenario is that he has central nervous system depression from barbiturates and is not breathing. This would explain the finding of his normal A—a difference.

CHAPTER 7

Regulation of O_2 and CO_2 in the Body and Acid/Base

So far in the book we've emphasized the role that O_2 plays in the production of energy in the body. We've reviewed how O_2 moves from the atmosphere to the alveolus and from the alveolus to the blood and the rest of the body. In parallel we've mentioned that CO_2 is a byproduct of tissue metabolism, and touched on the movement of CO_2 from the tissues to the lungs. If we've done our job of writing and explaining appropriately, the reason for the partial pressures of O_2 in the arterial and the venous systems should be clear by now. However, in all of this talk about moving O_2 and CO_2 however, we mentioned the pressures of CO_2 in the arterial (40 mmHg) and venous blood (45 mmHg) but didn't explain a particular reason for this. However, if you consider that the partial pressure of CO_2 in the atmosphere is close to 0 mmHg, it's just a little funky that the pressure of CO_2 in the blood is that much higher... don't you think? Well, guess what, CO_2 plays a *critical* role in regulating acid base metabolism in the body. However, before we delve into the fine details, let's take a big picture view of the problem so you can see why we need CO_2!

Note to reader: Acid/base is an incredibly complex topic. As we've mentioned previously it is not our goal to be a comprehensive textbook; rather we want to provide a bird's eye view of physiology to allow the reader to understand integrated physiology concepts and move on to more advanced textbooks. So keep that in mind as we move along.

WHAT ACTUALLY DRIVES VENTILATION?

Before we dive deep head first into the topic of acid/base, let's touch base on the regulation of ventilation. These two topics, although they might seem different, are actually part of the same overall mechanism, and a good understanding of the factors that regulate ventilation will help tremendously when trying to study acid/base disorders. So far in

Back to Basics in Physiology. DOI: http://dx.doi.org/10.1016/B978-0-12-801768-5.00007-1

this book we've discussed how CO_2 and O_2 pressures in the body are linked to ventilation. All other things being equal, more ventilation means higher O_2 and lower CO_2, while less ventilation means lower O_2 and higher CO_2. So if this is the case, then what regulates ventilation O_2 or CO_2? Well, how about both? In fact both CO_2 and O_2 regulate respiratory drive!

We'll spare you some of the neurological details, but suffice it to say that the impulse to breathe originates in the brainstem. It is an autonomous impulse that stimulates the contraction of the diaphragm, and as we saw in Chapter 3, contraction of the diaphragm brings air into the lungs. The more continuously the respiratory center fires, the more the diaphragm will contract and the more ventilation will take place. This process is the integration of three components (Figure 7.1): the sensors, the control center (brainstem), and the effector muscles (e.g., the diaphragm). Of these, the sensors can roughly be divided into central receptors and peripheral receptors. The central receptors sense CO_2, while the peripheral receptors sense O_2. Let's expand on this a little.

Central Chemoreceptors

These sensors respond to the increased acidity within the cerebral spinal fluid (CSF) (Figure 7.2A). The brain is separated from the blood by the blood—brain barrier, which is relatively impermeable to H^+ ions, but permeable to CO_2. Therefore CO_2 in the capillaries readily

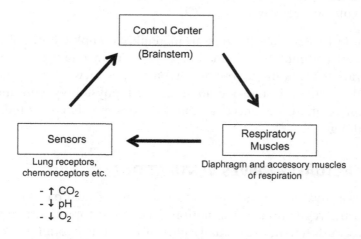

Figure 7.1 The three components of the regulation of ventilation. Central chemoreceptors sense the H^+ derived from the changes in CO_2. Carotid and aortic arch O_2 sensors respond to decreasing pressures of O_2, with an exponential increase in O_2 below 50 mmHg.

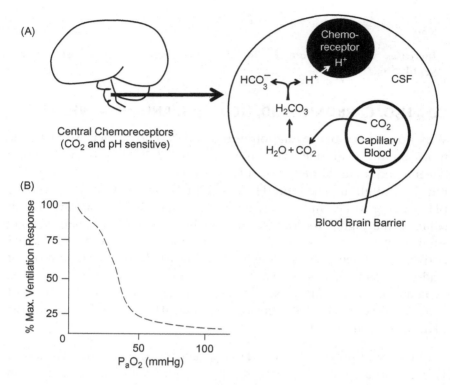

Figure 7.2 *(A) The central chemoreceptors are sensitive to the H^+ ions that are generated by the CO_2 that diffuses through the blood–brain barrier, and stimulates ventilation. (B) The stimulation of peripheral chemoreceptors by low O_2 is not a linear relationship. The lower the O_2 the higher the fold response is.*

diffuses into the CSF through the blood–brain barrier, and once inside the CSF reacts with H_2O, which will then generate H_2CO_3 and then dissociate into HCO_3^- and H^+. The H^+ ions are sensed by the central chemoreceptors. As CO_2 increases, the pH in the CSF decreases and stimulates ventilation. Similar to what happens with CO_2, although to a lesser degree, primary decreases in pH stimulate ventilation. The lower the pH, the more ventilation is going to be stimulated.

Peripheral Chemoreceptors

These sensors, located in the carotid bodies, are sensitive to CO_2, H^+, and O_2. But the most important one is O_2. As is logical, decreases in the pressure of O_2 increase ventilation. However, this only happens when O_2 is below 100 mmHg; as O_2 continues to decrease, the increase in ventilation is exponential (Figure 7.2B).

CO_2, H_2O, CARBONIC ACID, HCO_3^-, pH, AND pKa—WHAT?

Why do we need pH regulation in the body? Who cares if we're a little more on the acid side every now and then? Well, protein function depends on a stable pH! And, pH is a function of the amount of H^+ ions in a solution. The more H^+ ions that are in solution the lower the pH is going to be, and conversely as less and less H^+ ions are free in solution the higher the pH is going to be. A pH of 7 is neutral, any solution with a pH below 7 is an acid, and any solution with a pH above 7 is a base. According to the Bronsted–Lowry definition of acids and bases, acids release H^+ ions and bases bind H^+ ions. The behavior of acids and bases in solution depend on a multitude of factors, however, we will focus on two, the pH of the solution and the pKa of the compound.

As we said previously, the pH is dependent on the concentration of H^+ ions in a solution, so pH is a property of solutions. The pKa on the other hand is a property of the compounds themselves. The pKa is the level of pH where 50% of a particular compound is associated and 50% dissociated, therefore it is a measure of the strength of an acid in solution. What does this mean? Well, according to the Bronsted–Lowry definition of acids and bases, acids donate H^+ and bases accept H^+, and this is dependent on the pKa of each compound. So, a compound will behave as an acid (donate H^+) if placed in a solution that has a pH that is above the compound's pKa, and the same compound will behave as a base (bind H^+) if placed in a solution with a pH that is below the compound's pKa. In other words, below their pKa value, compounds hoard H^+ and above their pKa value compounds release H^+.

How is this relevant to our discussion? Let's say study compound HA! (This is the example that is always used in textbooks, so we'll play along.) Compound HA has a pKa of 7, which means that at a pH of 7, compound HA is 50% associated as HA, and 50% dissociated as H^+ and A^-.

$$HA \leftrightarrow H + A^-$$

So:

- At a pH that is higher than 7, HA will behave as an acid; that is, it will dissociate more and more, therefore adding H^+ to the medium to form $H + A^-$.
- At a pH that is LOWER than 7, HA will behave like a base; that is, it will associate more and more, therefore removing H^+ from the medium to form HA.

This concept is important to understand buffering in solution. A buffer is a compound that accepts or donates H^+ ions as needed in order to maintain the pH of a solution. Now think about what is going on with HA in each of the previous conditions. As the pH increases the compound actually starts releasing H^+, thereby decreasing pH, and as the pH decreases the compound actually starts binding to H^+, thereby increasing pH. Therefore a buffer helps maintain a stable pH! It takes a little and gives a little when needed. A compound buffers the best when it is close to its pKa value. This should make intuitive sense, because if what we're looking for in a buffer is the ability to both bind to and release H^+ ions, the best way to do that is if half of the compound is bound to H^+ (this can start releasing H^+ if needed) and half of the compound is dissociated from H^+ (this can start binding to H^+ if needed). In the human body, we consider a pH of 7.4 ± 0.05 normal so anywhere from 7.35 to 7.45 can be roughly considered normal. So that means that in order to buffer this system we would need a compound that has a pKa in this range! And we do, we actually have two systems that are very important:

- The H_2CO_3 system (carbonic acid): pKa of 6.1
- The H_2PO_4 system (dihydrogen phosphate): pKa of 6.8 to 7.2

As you can see, the pKa values of both these systems are relatively close to the pH value of 7.4 that we consider normal for the human body. Although the H_2PO_4 system has a pKa that's actually closer to the normal body pH than that of H_2CO_3, quantitatively the H_2CO_3 system is so large that it is the main buffer system in the human body.

Key

pH is a $-\log$ scale of H^+ concentrations. In simple terms the negative part of the $-\log$ means that if H^+ concentration goes up, pH goes down and if the H^+ concentration goes down, pH goes up. So low pH means more acidic and high pH means more alkaline.

The normal body pH of 7.4 requires a solution of 40 nanomoles of H^+ per liter. If we consider Na^+ concentrations in the body as a reference point, normal Na^+ concentrations are 140 mEq/L. So let's get some perspective on this; 140 mEq/L of Na^+ means that the concentration of Na^+ is 3.5 *million* times higher than the concentration of H^+ ions in the body. Yes, that is correct, 3.5 *million* times as much. This means that tiny amounts of H^+ can have a huge impact on pH. And why do we care about pH regulation in the body? Well, if we change the concentration of H^+ ions in the body, we can alter protein function. Hydrogen ions that are free in solution will attach to proteins and start degrading them. (This is the same principle that acid in the stomach uses to digest proteins. The hydrochloric acid that is secreted by the principle cells helps break down the peptide bonds that hold the amino acids together.) In order to keep the proteins in the body working as they should, it is extremely important to maintain a stable pH. There is a slight complication to all of this: Normal body metabolism produces about 100 mEq of H^+ ions a day (about 1 to 1.2 mEq/kg of body weight). This is called the endogenous acid load, which will be distributed in the 14 L of extracellular fluid This would raise the concentration of H^+ ions approximately 7 mEq per liter, which is 175,000 times the normal concentration of H^+ ions in the body, and the pH would be around 2! So clearly there has to be a way of keeping these H^+ ions from wreaking havoc in the body. The CO_2–Bicarbonate (HCO_3^-) buffer system does just that! As we mentioned before, the CO_2–Bicarbonate (HCO_3^-) buffer system can rapidly and very efficiently buffer changes in H^+ ion concentration and therefore maintain a stable pH. Let's take a look at how this is achieved.

HENDERSON–HASSELBALCH EQUATION

As we mentioned at the beginning of this chapter, in spite of the fact that CO_2 is a byproduct of metabolism, and some would think just a waste product, nature figured out a magnificent way to make use of it as a buffer to prevent pH changes in the body. The role of CO_2 as a buffer hinges on a single chemical reaction:

$$H_2O + CO_2 \leftrightarrow H_2CO_3 \leftrightarrow H^+ + HCO_3^-$$

This means that water and CO_2 will reversibly bind to form carbonic acid (H_2CO_3), which will in turn dissociate to form H^+ and bicarbonate (HCO_3^-). This reaction is a relatively slow reaction, but in the body, an enzyme called carbonic anhydrase (of which there are several subtypes in different tissues) speeds everything up. In fact carbonic anhydrase is one of the fastest enzymes in the human body! And what makes this reaction move in one direction or another? Well, since all the steps in this reaction are reversible, the concentration of the products on either end will drive the reaction forward. What? Think about it like this: Reactions that are fully reversible always attempt to maintain a balanced reaction. So if the compounds start building up on one end, they will actively start moving them toward the other end of the reaction. Having this in mind, the two you need to focus on are CO_2 and H^+. If the CO_2 starts building up, the reverse reaction will speed up and start turning that CO_2 into H^+ and HCO_3^-. Conversely, as H^+ starts building up, balance will be maintained by converting H_2CO_3 to H_2O and CO_2. Variations of CO_2 and H^+ that occur through this reaction will have a direct effect on the pH. The relationship between all these elements and the pH is summarized by the Henderson−Hasselbalch equation where:

$$pH = pKa + \log \frac{[HCO_3^-]}{0.03 \times PaCO_2}$$

According to the Henderson−Hasselbalch equation we can calculate the pH value for extracellular fluid. But, this is relatively complex, so let's simplify the relationship for you:

$$pH \propto \frac{HCO_3^-}{PaCO_2}$$

In other words, pH is directly related to HCO_3^- and inversely related to CO_2. This means that as HCO_3^- goes up, pH goes up and as the $PaCO_2$ goes up, pH goes down. So for all intents and purposes, CO_2 equates to acid and HCO_3^- is a base. If you add more acid or remove base, pH goes down, and if you add more base or remove acid, pH goes up. That's it!

Now, how can we apply this to our discussion about acid/base? Let's expand on the carbonic acid reaction a little. The initial reaction that we saw earlier does not explain the entire story:

$$H_2O + CO_2 \leftrightarrow H_2CO_3 \leftrightarrow H^+ + HCO_3^-$$

Consider this: If CO_2 starts to increase, this will drive the reaction toward the formation of H^+ and HCO_3^-. As the H^+ ions start accumulating, the pH starts to decrease, which isn't a good thing. How can the body turn the reaction around, so that instead of producing H^+ ions, it neutralizes them and then gets rid of them? Well, what if we decrease the CO_2? That would work, wouldn't it? Yes, it would! And how can we achieve that? By increasing ventilation and blowing off CO_2 into the atmosphere! This sequence of events fits perfectly with what we explained in our simplified version of the Henderson–Hasselbalch equation, where an increase in the $PaCO_2$ leads to a decrease in pH (acid), and a decrease in $PaCO_2$ leads to an increase in pH (base).

So the regulation of CO_2 allows for the maintenance of pH. This is why we need a $PaCO_2$ of 40 mmHg and an HCO_3^- of approximately 25 mEq/L. The $CO_2 - HCO_3^-$ buffer system allows for the buffering of an enormous amount of H^+ ions. However, you must keep in mind that buffering is not the same as eliminating them from the system. Buffering maintains the pH at a steady state in the face of varying concentrations of H^+ ions, but a buffering system has finite capabilities. In order to keep the body in working order, the excess acid or excess base must be excreted somehow. Pertinent to our discussion, the regulation of the pressure of CO_2 by the lungs plays a key role in the regulation of pH.

CHANGES IN CO_2 AND pH: RESPIRATORY ACID/BASE DISORDERS

There are four primary acid/base disorders:

- **Metabolic Acidosis**. A disorder that generally originates anywhere but the lungs and increases the concentration of H^+ ions in the extracellular fluid.
- **Metabolic Alkalosis**. A disorder that generally originates anywhere but the lungs and decreases the concentration of H^+ ions in the extracellular fluid.
- **Respiratory Acidosis**. A disorder that generally originates in the lung, which increases the $PaCO_2$, generally through a decrease in alveolar ventilation. Carbonic anhydrase will use the increased CO_2 to increase the production of carbonic acid (H_2CO_3), which then

dissociates and releases H^+ ions into the blood. Thus this would be the overall direction the reaction would be heading in:

$$CO_2 + H_2O \rightarrow H_2CO_3 \rightarrow H^+ + HCO_3^-$$

- **Respiratory Alkalosis**. A disorder that generally originates in the lung that decreases the $PaCO_2$, generally through increased alveolar ventilation. As the $PaCO_2$ decreases, H^+ ions will begin to bind HCO_3^- to make (H_2CO_3), which in turn will dissociate to H_2O and CO_2. Thus this would be the overall direction the reaction would be heading in:

$$CO_2 + H_2O \leftarrow H_2CO_3 \leftarrow H^+ + HCO_3^-$$

When we say primary, we mean where the problem starts. We divide acid/base disorders into respiratory problems and problems dealing with the rest of the body, which we call metabolic disorders. More often than not, metabolic disorders involve the kidney and acid that can't be excreted through the lungs. Respiratory disorders however, are directly linked to the amount of CO_2 in the body. Beyond the four primary acid/base disorders, acid/base gets progressively complex, because you can have a combination of multiple disorders that present at the same time! In the context of lung physiology the discussion regarding acid/base status centers around what the lungs are doing with the CO_2. So, considering that CO_2 has a direct impact on pH, a simple way to think about it is CO_2 = acid. Therefore if you want to decrease the amount of acid in the body, you increase ventilation and decrease CO_2, and if you want to increase the amount of acid in your body, you decrease ventilation and increase CO_2.

Key

Accumulating CO_2 in the body is converted into H^+. Therefore we can consider CO_2 an "acid."

General Principles of Mixed Disorders

As you can imagine, the body has a particular way of compensating the primary disorders that we described in the previous sections. Think about it like this: The body tends toward equilibrium, so if the primary disorder is metabolic, the compensation will be respiratory, and if the primary disorder is respiratory the compensation is metabolic. This means that if someone has a metabolic acidosis (there's an acid accumulating in the body), the body will try to compensate by getting rid of acid by decreasing the CO_2. The opposite is true for a metabolic alkalosis: the body will try to compensate by retaining CO_2. So if every primary disorder has a specific compensation, what do we call mixed disorders? Mixed disorders happen when you have two or more ongoing primary disorders. So, if you have two or more primary disorders, the body's ability to compensate can be hindered.

CLINICAL VIGNETTES

A healthy 24-year-old woman decided to climb Mt. Everest. She prepared arduously for months before her trip and finally arrived in Nepal for the trek. She arrived at base camp, which is roughly 18,000 ft above sea level (5,400 m). The atmospheric pressure at base camp was around 400 mmHg (as opposed to the 760 mmHg at sea level). Her respiratory rate (RR) at home was around 10 breaths per minute. However at base camp her RR increased to 17 breaths per minute. One of the guides gave her acetazolamide (which is a carbonic anhydrase inhibitor) to help with the acclimatization.

1. Why would acetazolamide help with acclimatization to high altitude?
   A. It increases the excretion of CO_2.
   B. It decreases the excretion of CO_2.
   C. It has no effect on acclimatization.

Answer: B. Among the many effects that acetazolamide has to favor acclimatization to high altitude, one is the decrease of the inhibition of the respiratory drive stimulated by the low O_2. What? As the atmospheric pressure decreases, the amount of O_2 that our patient is breathing decreases as well (from 760 mmHg to 400 mmHg the partial pressure drop in O_2 goes from 150 mmHg to 74 mmHg—almost half!). The drop in O_2 would stimulate ventilation, but as we also saw previously, increased ventilation will decrease CO_2. Decreases in CO_2 will blunt the increased ventilation response, so basically our patient won't be able to breathe enough O_2 because her CO_2 is too low! By giving acetazolamide, which blocks carbonic anhydrase, we generate a metabolic acidosis by blocking HCO_3^- reabsorption by the kidney, thereby opposing the respiratory alkalosis generated by the low O_2. This allows for a continued and more robust hyperventilation response triggered by the low O_2.

Clinical Recognition: Signs and Symptoms of Respiratory Distress and Their Physiologic Basis

As we've said over and over again in this book, the role of the ventilation system is to move air in and out of the alveoli in order to maintain a high P_AO_2 and a relatively low P_ACO_2 to allow for exchange with the blood. The role of the cardiovascular system is to carry gases to and from capillaries where they can be exchanged by diffusion. And we've reviewed some equations that allow us to quantify these phenomena.

In order to measure the amount of air going in and out of the alveoli, we have the alveolar ventilation equation:

$$V_A = RR \times (V_T - V_D)$$

In order to calculate the amount of O_2 in the alveoli, we have the alveolar gas equation:

$$P_AO_2 = FiO_2(P_{ATM} - P_{H_2O}) - P_aCO_2/RQ$$

In order to calculate the quantity of O_2 in the blood, we have the delivery of O_2 equation:

$$DO_2 = CO \times Hb \, (g/dL) \times 1.34 \, mL \times Saturation$$

And in order to quantify how much of the delivered O_2 is being consumed, we have the O_2 consumption equation:

$$O_2 \; consumption = Arterial \; O_2 \; content - Venous \; O_2 \; content$$

Although it may not seem like it at first, all of these equations are linked to one another since they all describe the steps required in getting O_2 to the cells and CO_2 out of the cells of the body. Think about it like this: If we decrease alveolar ventilation, then the P_AO_2 is going to decrease. If the P_AO_2 decreases, then the O_2 content will decrease, and O_2 delivery will decrease as well! These equations can be used to readily explain respiratory failure in its simplest form.

Back to Basics in Physiology. DOI: http://dx.doi.org/10.1016/B978-0-12-801768-5.00008-3

But the question remains, how do you recognize this clinically? Are there simple clinical signs to help you in your understanding? In this chapter we will cover a general overview of respiratory failure, followed by clinical signs that will allow you to recognize it as it's happening. In the final chapter, we will ultimately walk you through clinical cases, which should hopefully put everything together for you.

CAUSES OF RESPIRATORY FAILURE

If you look up any reference text, you may see respiratory failure broken down into two types. Type I respiratory failure is classically described as failure of oxygenation of the blood. Type II failure is described as failure to remove carbon dioxide from the blood. These lists further break down respiratory failure into additional subtypes, including V/Q mismatch, shunt, and so on. Much of this we've already addressed in previous chapters looking at things functionally. Ultimately though, *respiratory failure can be distilled rather simply into inadequate ventilation or inadequate oxygenation.* In some cases, these are *not* mutually exclusive! Obviously, if you simply stopped breathing, you would not ventilate, which means you subsequently would not oxygenate.

Rather than make you memorize a long list of causes of respiratory failure, we're going to attempt a convenient method to think about it using some very basic anatomy and what you've learned so far. Our goal is that you *understand* this stuff and not just memorize it!

First, we'll include a very small review of anatomy. Take a look at Figure 8.1. We can very generally separate the relevant anatomy into two segments: one that is above the clavicle and one that is below the clavicle. Up until now we've focused mainly on the parts below the clavicle known as the intrathoracic, or "inside the thorax," components. This is because it is here that the main components that make up the lung and conducting airways lie. These are all subject to the *negative* pressure as discussed in Chapters 3 and 4. However, if we're looking at areas where things can go wrong, we must look at everything from the air itself all the way down to the diaphragm! Here you see that the air passes through the nose/mouth, nasopharynx/oropharynx, and makes its way down into the larynx passed the glottis

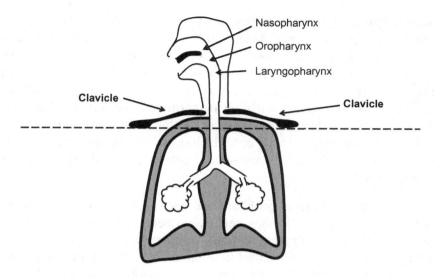

Figure 8.1 Causes of respiratory failure can be roughly subdivided into those that have an origin above the clavicle (dotted line) and those below the clavicle. Problems that are originated above the clavicle are generally ventilation issues (not enough air coming in), whereas problems below the clavicle can be problems ventilating and/or problems oxygenating the blood.

following the pressure gradient into the relatively negative pressure environment of the thorax below the clavicle. It then gets its way down into the lower parts the lung and ultimately to the alveoli, where gas exchange occurs with the red blood cells in the alveolar–capillary units. If you recall the diagrams with the balloons in jars in Chapter 3, the thorax is the jar. Now then, on to problems ventilating, the easiest to understand.

Problems Ventilating

When you think of ventilation issues, think about problems moving air in and out. That's it. Any disease process that interferes with the movement of air in and out of the lungs is a ventilation issue. In this regard, basically any problem above the clavicle is a problem ventilating because these "above the clavicle problems" generally impede airflow down to the lungs. It's that simple! Some causes of "above the clavicle" ventilation issues are a lack of respiratory drive from the brain, such as seizure, opiate intoxication, an obstruction in the nose, mouth, pharynx, or larynx. All of these problems result from whole-lung decreases in ventilation—the wind pipe isn't bringing air in! Now, you can also have ventilation problems below the clavicle, and we can also subdivide these anatomically as we make our way down the respiratory tree. There can be an infection and swelling at the level of the trachea (tracheitis), bronchi (bronchitis), bronchioles (bronchiolitis), and alveoli (pneumonia, ARDS) that impair or interfere with the movement of air in and out of the lungs. Similarly, you can have problems with obstruction not caused by infection, but caused by inflammatory responses that cause swelling such as asthma. Additionally, there can be diseases where the organization of the lower airways themselves gets destroyed such as in emphysema.

In general, ventilation problems are ones that can be improved by increasing ventilation. Remember that alveolar ventilation is determined by respiratory rate and tidal volume. It is usually a primary defect in one of these two areas that cause Type II respiratory failure. As such, this type of respiratory failure is typically characterized by an elevated partial pressure of CO_2 in the blood.

Problems Oxygenating

Hypoxemia, or low O_2 in the blood, is a whole other ball game. Certainly if you can't move any air into the lungs, as we saw with hypoventilation, you're not going to oxygenate blood very effectively. But as previously explained if you're not ventilating, CO_2, is also going to increase within the blood (Figure 8.2). So in broad

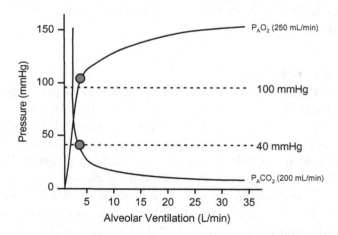

Figure 8.2 Relationship between alveolar ventilation (V_A) and pressures of O_2 and CO_2 in the alveoli. As V_A increases, O_2 increases and CO_2 decreases. As V_A decreases, O_2 decreases and CO_2 increases.

strokes, when we refer to problems with oxygenation and thus hypoxemia, we're talking about problems getting O_2 into the blood. This could be due to hypoventilation of course, but also other factors. For example, it could be due to the air itself. If a totally healthy person were to be placed in an environment where the air he or she was breathing in contained little-to-no oxygen (e.g., high at the top of Mount Everest), ventilation could be fantastic, but what good would it do if the air circulating in the lungs doesn't contain enough oxygen? Well, your CO_2 will be exhaled just fine, in fact it would be low as you increased ventilation to make up for low blood oxygen levels. If oxygen levels in the air dropped too low of course, you wouldn't be breathing much at all since you'd likely be dead!

There can be problems at the level of the alveolar capillary unit that result primarily in hypoxemia without necessarily a problem ventilating or removing CO_2 from the blood. For example, there can be a barrier to diffusion. If you were to increase the distance between the alveoli and capillary (as in the case of pulmonary edema, or due to some sort of inflammatory disease process that causes scarring at the interface as in the case of pulmonary fibrosis) you can have problems oxygenating.

These are types of respiratory failure that occur because the distance the oxygen needs to travel to get into the blood is too long for efficient diffusion. Because CO_2 has an easier time diffusing out than O_2 has getting in (due in large part to differences in their solubility), diffusion problems tend to be one of low partial pressures of oxygen in the blood, but normal levels of carbon dioxide. These diseases will also become more apparent when there's an increase in the demand for O_2.

You could also have problems that result from an imbalance between the capillary blood flow and ventilation of the alveoli, otherwise known as V/Q mismatch as we saw in Chapter 6. In these instances, there can be too much blood flow in areas that are ventilating poorly or not enough blood flow in areas that are ventilating well. The end result is an overall problem with oxygenation. (Remember our discussion on the A—a difference from Chapter 6.)

Clinical Correlate

Pulmonary Embolism

A pulmonary embolism (PE) is the term used when a blood clot travels to the pulmonary circulation and gets lodged in one of the segments of the pulmonary artery. During a PE a V/Q mismatch is created because part of the pulmonary circulation is plugged up with a clot. With a PE, ventilation is fine (there's nothing opposing air going in and out of the lungs). The problem is the blood. In general, there is a lot of reserve in the respiratory system, so if a few small capillaries get plugged up, no big deal. The clot forces blood flow to be diverted toward all the "unplugged" lung vasculature. If it's a small clot you will see little evidence clinically as the body continues to oxygenate and remove carbon dioxide without a problem. (In healthy people the body has a *huge* physiologic reserve).

However, if the clot takes up a large number of capillaries all at once, or gets plugged up in a larger vessel, there could be problems. As the blood is diverted to pulmonary vessels that are not plugged up with a clot, eventually it becomes a problem of surface area, where there is just too much blood passing through the unobstructed blood vessels for an adequate exchange to take place (essentially we're back to a $Q > V$ scenario) and there is insufficient oxygenation of the blood going through. This is basically why you see hypoxemia with PE!

You may have heard of a medical emergency called a "saddle embolus." This is dangerous because it is a clot that sits at the bifurcation of the main pulmonary artery where deoxygenated blood leaves the heart and where it splits into the right left main pulmonary arteries going to off to the right and left lungs. This is such an emergency because you're blocking blood flow to both lungs!

An extreme example of the V/Q mismatch scenario is a shunt. A shunt is a condition in which deoxygenated blood bypasses the lungs entirely. This can be something structural like a hole in the heart (ASD/VSD) where, if there is sufficient pressure on the right side, deoxygenated blood gets pushed from the right side of the heart to the left, bypassing the lungs. Or it could be a situation in which blood that is going through the lungs is not getting oxygenated at all, thereby "functionally," if not physically, bypassing the lungs. So as the deoxygenated shunted blood mixes in with the oxygenated blood, the amount of O_2 being carried by blood can only decrease.

These two things, shunt and V/Q mismatch, are frequently a source of confusion for medical professionals. So let's give another example to reinforce the difference. Let's pretend now that a patient of ours has an infection, and a huge mucus blob plugs up one of the larger branches of the lung, effectively reducing ventilation of that entire segment of the lung to zero. You might think that this would primarily be a Type II cause of respiratory failure because the problem is ventilating, but remember the CO_2 has an easier time getting out than O_2 has getting in. The area downstream of the plug will no longer be ventilating, and thus the pulmonary vessels will constrict to redistribute blood elsewhere in the lung where ventilation *is* taking place. Normally vessels dilate if the tissues they supply aren't getting enough oxygen, but in the lungs it's the opposite! If you think about it, this is a pretty good safety mechanism, because vasoconstriction in areas with poor oxygenation limits "wasting" blood flow on bad areas of the lung and acts to maximize the use of the good areas. This is a wonderful mechanism when only a small portion of the lung has low O_2. If the portion of the lung that has low O_2 gets too large, however (like in our example), the blood flow that gets redistributed to the other segments that are ventilating is just too much for those segments to handle. So in those segments Q will be a lot bigger than V. Because CO_2 gets out easier than O_2 gets in, the partial pressure of CO_2 in the blood might be normal, but there's not enough time to fully oxygenate all the blood flowing through, and so you will see a low partial pressure of oxygen within the blood. Because a shunt is, in a way, simply an extreme form of V/Q mismatching, in the hospital setting you may find that the two terms are (incorrectly) used interchangeably. For example, a doctor may say, "oh this patient is shunting" when explaining why they have a low oxyhemoglobin saturation on pulse oximetry. This is not accurate, strictly speaking, since a shunt is a complete bypass of the alveoli. Hopefully, you'll know better than to correct them though!

If any of this confuses you, don't worry. We'll go over it within the context of some clinical examples in the next chapter, but now that you have a general overview of what respiratory failure is, let's talk about what it looks like!

WAYS THAT VENTILATION CAN BE IMPROVED

Throughout the book we've discussed what normal ventilation looks like, and what it achieves. Now if we're having problems ventilating, what do you think we can do to correct these issues? As we move through the rest of this chapter consider the V_A formula once again:

$$V_A = RR \times (V_T - V_D)$$

As we saw in Chapter 4 there are two ways to increase V_A: increasing the respiratory rate and/or increasing the V_T. You may recall that out of the two, increasing V_T has a larger effect on V_A than RR alone. However, increasing the V_T means increasing the work of breathing. Think about it: In order to breathe in more air and thus increase the V_T, you need to decrease the pressure inside the thorax by a lot, which takes more energy. Therefore the simplest method to improve ventilation is to increase respiratory rate. In adults, an increased respiratory rate ($>15-20$ breaths per minute) is indicative of significant respiratory disease. In children this number varies. A newborn's respiratory rate is somewhere between 40 and 60 breaths/min. As babies get older, and their chest wall becomes more rigid, allowing the generation of larger tidal volumes, this number falls closer to the 40s, 30s, 20s, and eventually to the teens, which is a normal respiratory rate for an adult. If you had an adult patient breathing 60 times/min, you'd better get them help and fast! That would be an absolute emergency. However, a newborn baby? Par for the course. It is not uncommon for newborn infants sick with a respiratory infection such as bronchiolitis to breathing in the 70s and 80s, sometimes even exceeding 100 breaths/min! In medical parlance, increased respiratory rate is called *tachypnea*. A patient that is breathing fast is said to be *tachypneic*. A patient that is tachypneic is in distress and needs treatment for the underlying cause of his condition. So, increased work of breathing and tachypnea are clinical signs associated with respiratory failure. Let's take a look at some other clinical signs of respiratory failure that we can use to evaluate our patients.

CLINICAL SIGNS OF RESPIRATORY FAILURE

In clinical practice, your patient will not walk up to you and say, "My V_T decreased, probably because I have asthma and I just smoked a cigarette." Rather, you'll get called because a patient just came in who can't breathe, and it's your job to figure out what's going on. You've probably heard the common adage: "90% of conditions can be diagnosed with a good history and physical." Respiratory failure is not the exception, but you have to know what to look for. Many of the clinical signs discussed here fall under the general umbrella of "increased work of breathing" or "increased respiratory effort," which you may find written in many a medical chart. The reason behind these clinical manifestations is varied, but usually it boils down to trying to improve ventilation for the purposes of either getting more oxygen in, or getting more CO_2 out. So let's go through these signs starting from the top and working our way down.

Nasal Flaring

Compensation for poor ventilation at the level of the nasopharynx is most commonly is seen as "nasal flaring," in which the nasal openings widen. Because there is a great deal of airway resistance in the nose (they're two small holes, after all), adults that are having problems ventilating can choose to just breathe through their mouth. Infants, however, are obligate nose breathers up until 3 to 6 months of age. This facilitates suckling at the breast and not dying. They frequently are only seen breathing through their mouths when crying. As a result, when infants are having significant problems ventilating, it is not at all uncommon to see them flaring their nostrils as they inhale. By making the nostrils bigger, resistance decreases, and air flows more easily into the nasopharynx.

Clinical Correlate

Choanal Atresia

Choanae is from the Greek word meaning "funnel." It is represents the back of the nose and is the connection between the nose and the nasopharynx. During gestation *in utero*, there is a membrane between the nose and the nasopharynx that disappears before birth; however, in patients with choanal atresia this does not happen and the membrane remains. As we just said, newborns are obligate nose breathers, so you can imagine their dilemma if they're born with this problem! This congenital abnormality is often diagnosed before the infant leaves the hospital. If it affects both the right and left nasal cavities, you may find a patient who is blue (cyanotic) at rest or especially when feeding but appears bright pink and healthy when crying!

Stertor

Anyone who has ever heard someone snore has heard stertor before. Stertor is the noise that results from vibration of the pharyngeal tissues (nasopharynx, oropharynx, soft palate) due to significant upper respiratory obstruction and subsequent turbulent airflow downstream in the upper airway. This is a noise heard only on inspiration. This is an "above the clavicle" problem and so ventilation may be a significant concern in patients who are heard "snoring." Always keep a high level of suspicion when you hear stertor, as it could very well mean that your patient is having a difficult time breathing.

Clinical Correlate

Obstructive Sleep Apnea

Patients with severe snoring problems sometimes suffer from so much obstruction that they actually don't breathe at all for periods of time before suddenly taking sudden gasping breaths. This is a medical condition known as obstructive sleep apnea (OSA), and is often due to relaxation of the pharyngeal wall muscles during sleep. It can be complicated by enlarged tonsils and adenoids as well. Patients with long-standing obstructive sleep apnea can actually begin to retain CO_2 and develop heart problems as well, secondary to chronic ventilatory problems.

Cough

Perhaps one of the simplest, but often most overlooked, signs that could indicate problems ventilating is cough. The purpose of cough is to expel irritants and obstructions in the respiratory tree. It is a vital part of the body's natural defense against pollutants/microbes as well as a means to help ensure adequate tidal volume in the case of airway obstruction from mucous/debris plugging the airways. Coughing is stimulated by various "irritant" or chemical receptors in the larger airways and near glottis. There are also stretch receptors in the lower lungs as well that can stimulate cough.

Coughing occurs as a result of the glottis opening and taking in a brief, reasonably large inhalation followed shortly by the glottis closing. The diaphragm relaxes and the chest muscles, abdominal muscles, and even the pelvic muscles forcefully and rapidly contract, rapidly increasing the intrathoracic pressure as you try and exhale against a closed glottis. The glottis then suddenly opens releasing the flood gates as a burst of high-velocity air jettisons airway contents upward and out of the trachea and larynx. The mucous down in the lower airways also mobilizes as the smaller airways compress like a tube of toothpaste, squeezing the mucous more toward the larger airways, where subsequent coughing or the mucociliary escalator can help to remove the contents from the lungs.

Clinical Correlate

Cough

Cough in patients should be taken seriously. While most commonly it may simply be due to a common viral upper respiratory tract infection (URI—a cold), it is often the first sign of more ominous disease. While taking care of patients, one of your primary jobs is to recognize when it is *not* something common, and so additional history may be helpful to screen when a cough is not just simply a cold. In children, chronic cough, especially nighttime cough in the absence of runny nose (*rhinorrhea*) or other symptoms suggestive of a cold should prompt a clinician to think about undiagnosed asthma, which frequently presents with only coughing, and often coughing that wakes up a patient at night. If a fever is present with cough in the absence of *rhinorrhea* or nasal congestion, it should prompt consideration into possible nonviral infectious causes. Bacterial pneumonia commonly presents with cough, fever, and fast breathing in absence of any nasal symptoms, sore throat, and such.

Abdominal pain is a frequent complaint as well. This is because the phrenic nerve, which innervates the diaphragm, also meets up with several sensory nerves from the abdomen before heading back to the cervical spine. Lower lung pneumonias especially can irritate the diaphragm and the phrenic nerve, making the patient localize pain to the abdomen. Similarly, irritation of diaphragm frequently results in a decreased appetite as well. In a patient who has neurological deficits (cerebral palsy, stroke victims, etc.) and limited mobility, you should consider aspiration pneumonia (i.e., pneumonia caused by aspiration of "secondary" abdominal contents or "primary" mouth secretions/food). Chronic fever and cough in a patient with a history of weight loss, history of travel to developing countries, or history of HIV should prompt consideration of possible tuberculosis. Long-standing cough in a patient with a history of smoking should prompt consideration of COPD (e.g., emphysema) or even lung cancer.

Stridor

Stridor is a problem typically seen in children, but it can also be seen in adults. Indicative of another typically "above the diaphragm" problem, it is a harsh, high-pitched, musical sound heard most commonly on *inspiration*. It can be heard during expiration as well (not to be confused with wheezing, as we'll see shortly). Stridor results from vibration of the upper airways due to upper airway obstruction that is typically at or just below the level of the vocal cords (glottis). When the pleural pressure decreases to bring air in, in normal conditions the upper airway collapses ever so slightly. If there is swelling of the airway walls causing narrowing or an obstruction within that portion of the airway, the small collapse can make a bad problem much worse. Common causes can include foreign body aspiration, swelling due to infection, airway trauma, or vocal cord dysfunction.

Clinical Correlate

Croup

Croup, or laryngotracheobronchitis (great word, no?) is an inflammation of the upper airways caused most commonly by viral infection. It is seen commonly in children between 6 months and 6 years of age. In fact, one of the most common ways children would die from measles was due to this complication! With mild croup, you typically hear inspiratory stridor with agitation (the faster you're breathing, the faster the air is moving and the more turbulent the flow). More severe obstruction can result in stridor heard at rest (you don't even need a stethoscope!). More severe still, stridor can be heard during expiration *as well as* inspiration, known as biphasic stridor. This is because the obstruction is so severe, it doesn't even need that mild compression of the airway during inspiration; the airway can even be distended during exhalation and you can hear it because the airway is severely narrowed (think of whistling, it can be done inhaling *or* exhaling). Increased work of breathing associated with stridor is a particularly worrying sign because it's a sign that ventilation is being impaired. If ventilation is impaired significantly, then narrowing of the airway is quite severe (imagine running while breathing through a straw). If there is hypoxemia associated with croup, this is an ominous sign! This implies that ventilation is so impaired that the body can't even oxygenate. In this setting the stridor may not even be audible since the obstruction is so severe that air movement is stifled almost entirely.

Wheezing

Wheezing is not unlike stridor, in that it too results from obstruction of the airway. However, this obstruction is uniformly involving the lower airways, not the upper! It is most commonly heard in patients with asthma and is heard because of fluttering of the walls of the intrathoracic airways at the site of obstruction. Because of this, you might be able to surmise that wheezing occurs primarily on *exhalation* when in milder forms. During exhalation, the intrathoracic airways narrow ever so slightly. Again, the same rules apply as with stridor, but in reverse. If more severe in nature, it can be heard during *inspiration* as well, but if very severe obstruction occurs, you may hear next to no air movement!

Clinical Correlate

Asthma and Wheezing

Asthma is one of the most common causes of recurrent wheezing. Wheezing is clinically significant! It typically implies that peak expiratory flow is less than 50% of the normal value! It is important to note, however, that all that wheezes is not asthma! One of the most common respiratory infections seen in infants is viral bronchiolitis. This is inflammation of the bronchioles (which are by definition lower airways). This inflammation produces swelling, mucus, and sloughing off of dead airway epithelium that makes its way down into the lower lung. Because children's airways are narrower even when well, relatively small amounts of swelling can produce this sign. In general the differential is actually very broad. When paired with other atopic conditions such as eczema it is easier to diagnose, but wheezing is so common in infants that, when recurrent, it has led to its own diagnosis called "reactive airway disease." In general, this is a bit of a misnomer and has fallen out of favor as wheezing with viral infection is so common.

Crackles

Crackles are another physical exam finding common in patients of all ages. They have been described as discontinuous adventitious breath sounds, which really doesn't help that much in understanding what they sound like. To know what they sound like, aside from examining a patient who has them, they can be nicely approximated by rubbing two hair strands between your fingers next to your ear. The noise is actually being made from the snapping open of alveoli, and are thus heard during and up to the end of inspiration. They are frequently heard in various types of lower respiratory tract disease or heart disease (with pulmonary edema). They come in a few flavors, some sounding finer and some sounding coarser, depending on the size of the airway from which they are produced. Most commonly, they can be heard in pneumonia, heart failure, asthma, bronchiolitis, and the like, but they can also be heard in normal patients, especially first thing in the morning.

PUTTING IT ALL IN CONTEXT

The key as a young medical care provider is to promptly recognize in very broad strokes what could be causing the problem, and then to quickly and decisively take the next steps to prevent the patient's clinical deterioration. Taking the clinical signs we describe as a starting point, if we think of diseases that affect ventilation in an organized manner, we will be able to understand at what level the problem is located, and therefore what the treatment should be aimed at. For example, if you have a ventilation problem, you improve ventilation; if you have an oxygenation problem, you work to try to improve oxygenation! By paying attention to these clinical signs and combining them with what you learn from the rest of your history, physical exam, and additional diagnostic testing, you can pinpoint the exact cause of these problems and try and treat the source! Next let's take a look at various clinical scenarios to put all our knowledge to the test.

CHAPTER 9

Clinical Integration

So you've gotten through all the pathophysiology and you're feeling good. You feel like you're ready to go and save lives. Well, here's your chance! In this chapter we're going to focus on clinical cases, even though your knowledge base may not be at the level yet where you will understand the individual medical therapies/technologies in any great detail, the point is that you will understand the concept behind their use. After all, many of medicine's most serendipitous and useful innovations have come from common sense applications of pathophysiology distilled to its simplest form. So don't worry if you don't know individual medicine names or mechanisms of action. Don't worry if you've never heard of certain respiratory devices (the appendix details them if you'd like some reference). Don't worry if you've never taken a CPR course. If you've read this book, you'll be adequately prepared to tackle this chapter.

So our first step whenever we face a patient that is having difficulty breathing, is to identify the problem and stabilize our patient, and to do this as swiftly as possible. To do that, we need to identify exactly where the problem is occurring. Throughout the course of this book, we've identified three potential problem areas:

- Getting air into the lungs and down to the alveoli
- Getting the oxygen from the alveoli into the blood as it travels through pulmonary blood vessels
- Carrying sufficient oxygen in the blood to supply the cells that need it most

In each of the following cases, we're going to apply what we've learned throughout the book by trying to distill all this pathophysiology into the most pressing questions we need to ask ourselves while evaluating a patient in respiratory distress. To figure out which one of

Back to Basics in Physiology. DOI: http://dx.doi.org/10.1016/B978-0-12-801768-5.00009-5

the functions just outlined is affected we will ask these three questions when evaluating each clinical scenario:

1. *Is the patient breathing?*
2. *Is the O_2 getting from the alveoli into the blood?*
3. *Does the patient have an adequate capacity for transport and delivery of O_2 assuming we can get more O_2 into the blood?*

and then conclude with the final question:

Distilled down, what's the patient's basic problem, and what are you going to do to stabilize it?

Now, without further adieu...

CASE #1

You're working the late-night shift in the Emergency Room and you go into an examining room to see a 17-year-old patient whose chief complaint is seizures. In talking with the family, you learn that this patient has a long-standing history of epilepsy, and he ran out of his medication several days ago but has yet to get his prescription refilled. He had a prolonged seizure at home, which resolved spontaneously. Parents noted that he seemed to turn purple/blue around the lips during the time of the event. Aside from his history of epilepsy, he has otherwise been healthy and has had regular checkups with his pediatrician and neurologist. He arrived via ambulance after the parents called 911, and the EMS team placed an IV and gave him some antiseizure medicine called lorazepam. On exam the patient is initially very sleepy and difficult to arouse. Upon trying to examine the patient he begins to have another seizure. He begins flailing his arms and legs and periodically stiffening them, staring off into space with a fixed gaze. You notice that his lips are beginning to turn a purplish/blue.

What are your first steps? What are you going to do right now? The mother is looking at you saying, Dr. [Insert Name Here], what are you going to do? YOU NEED TO ACT!

Since this is our first case, the first thing you're going to need to do is NOT panic! I know that was a lot of capital letters, but you've got this! In order to identify what we need to do first, we need to get our medical priorities straight. Way back in Chapter 1 we explained what kills you the quickest, and it's not seizure, it's lack of oxygen. This patient has had seizures his entire life, so another seizure probably isn't going to kill him. Yes, you need to treat the seizure, but first you need to ask yourself, why is my patient blue? Assuming this patient doesn't walk around blue all the time, we can surmise why he might be blue, no? We know that the blood likely does not have an appropriate amount of oxygenated hemoglobin at this very moment in time! So we now know what the most dangerous problem facing this young man is—he's not breathing!

As we will for all our cases, let's start off with our three questions:

1. *Is he breathing? In other words, is the air getting to where it needs to be?*
   Our first steps should be to swiftly determine whether air is getting to where it needs to be. Is air, specifically oxygen within the air, getting to his cells at this very moment? If not, we need to fix that. You look at the patient's chest and see that he is not breathing. He has essentially no respiratory effort. We said that problems above the clavicle (e.g., brain) tend to be of a ventilatory nature, and this is no exception.

2. *Is the oxygen getting appropriately from the lungs into the blood?*
   Since his lips and mucous membranes are blue (a sign of poor O_2 carrying by hemoglobin), probably not. Since we've established that he's not breathing, we don't really have to worry about V/Q mismatch and the like. We can safely assume he's not oxygenating because he's not ventilating, especially since he was pink a few seconds prior to the seizure. If he wasn't pink or had any other associated issues, we would first fix the ventilation and then start thinking about any other underlying problems such as a V/Q mismatch.

3. *Does the patient have an adequate capacity for transport and delivery of O_2 assuming we can get more O_2 into the blood?*
   We know that the delivery of oxygen has two components, right? Recall from Chapter 5: DO_2 (delivery of O_2) = CO (cardiac output) $\times$ O_2 content. What about the patient's cardiac output? Remember that the heart is a hungry organ when it comes to O_2 consumption, since it needs O_2 to generate ATP to keep beating, so be sure to check a pulse! If there's no pulse, that is a very good indication that either the low O_2 content has caused major damage to the hungry heart muscle, or there was a problem that originated in the

heart in the first place (in fact, CPR teaching as of the time of this book actually recommends checking a pulse first thing if you find a patient unconscious and are unsure of the cause). A pulse also tells you that if you can bring the O_2 content back up, delivery of O_2 should return to appropriate levels. While thinking about this, you check a pulse in our 17-year-old patient and breathe a small sigh of relief to find that it is 120 bpm and strong. The mom volunteers that this patient has no history of any heart problems.

What about his O_2 carrying capacity? Remember, O_2 carrying capacity is a function of the amount of hemoglobin in the blood and how much O_2 that hemoglobin is carrying aka saturation. At no point prior to this patient stopping breathing did you appreciate pallor, and his parents deny any (pallor could be a sign of anemia). He's male so isn't menstruating. It's unlikely at this point that an inadequate quantity or quality of hemoglobin is related to this problem.

Distilled down, what's the patient's basic problem, and what are you going to do to stabilize it?

He's not ventilating because he's not breathing. He's not breathing because of a problem "above the clavicle" (i.e., the brain). If his respiratory rate is 0, then his minute ventilation ($RR \times V_T$) is 0! This is why he's not oxygenating. So while you eventually need to treat the seizure, you first need to get the patient to ventilate.

In order to do that you will need help. So you call for help from your medical team (nurses, etc.). You ask a medical team member to bring you a facemask attached to a ventilatory bag, commonly called bag valve mask (BVM) or Ambu® bag, and you ask them to do it STAT. You could not simply place a non-rebreather mask or nasal cannula on this patient and walk away! (These devices only increase the amount of O_2 in the air, but the patient still needs to ventilate by himself.) This is a patient whose primary problem is that he is not ventilating. If he's not ventilating, it doesn't matter how high the oxygen concentration is if it doesn't get inhaled!

The first medical team member returns with the BVM and hands it to you. You start "bagging" the patient, squeezing the BVM. You ensure there is an adequate seal of the facemask around the nose and mouth. You see good chest rise and don't hear any air leak so you can be pretty sure all that gas you are pumping is going into the lungs. Now what?

Well, it would be a good idea to assess beyond chest rise and checking facemask seal that the patient is actually benefitting from your intervention! You don't want to stop bagging the patient, because he is still not breathing on his own. You see condensation in the mask, which is a good sign that the patient is probably exhaling humidified CO_2-rich air, and thus that his ventilation has improved. And even though it's not very easy to see his lips through the mask, you get the impression he seems a little pinker. A simple thing to do would be to ask one of the medical team members to attach a pulse oximeter (which measures the O_2 saturation of hemoglobin) on the patient's finger so you can better assess his response your bagging. Aren't you glad you asked for that help first thing?

The pulse oximeter is attached and hooked up to the monitor. You see a pulse rate of 115, there is a regular waveform and his oxygen saturation has come up from the 70 s and 80 s to 100%.

Great, job well done, right? Time to go home? No! You've simply stabilized this patient's ventilation and thus oxygenation problem. If you stop "bagging" the patient he'll stop ventilating again, so this is hardly a long-term fix. You now need to get to the business of providing treatment for the underlying cause. In this case, it would be fairly safe to surmise that the most likely cause is a seizure. And so you want to promptly get to the business of fixing that.

You give the patient two separate boluses of lorazepam. After the second bolus, the flailing and stiffening stops, after several more seconds you notice the patient begins to breathe on his own. Good job!

Here, you would move into the post-resuscitation phase of things. You would need to determine whether or not the patient remains coherent enough to protect his own airway, whether or not there is any sign of blood in the mouth (from possibly biting his tongue), or signs of vomiting that could imply risk for worsening respiratory problems assuming he aspirated some of it. If he is unable to do those things, or you worry about aspiration and a deterioration of his condition, you could consider intubating this patient with an oral endotracheal tube. Alternatively, if he's protecting his airway, you could monitor him very closely over the next few hours clinically and with the aid of continuous pulse oximetry and hold off on intubating him. If he stabilizes, you might want to call his neurologist as well to get recommendations on additional antiseizure medicines you may want to give this patient, since lorazepam is a fairly short-acting acute remedy.

CASE #2

A 2-year-old previously healthy boy presents to your office with a 2-day history of nasal congestion, rhinorrhea, and a low-grade fever to 101°F. One day prior, he began having a cough that since last night has sounded progressively more "barky" in nature. On exam, you find the child to be breathing fast, around 48 times per minute, with some suprasternal retractions and you appreciate stridor on inspiration.

What do you do next?

Well, as usual, you do a quick assessment with respect to the patient's color and check his pulses. You want to appreciate his level of alertness and general appearance. You see that he is nice and pink. He has good strong pulses with a heart rate of 140 bpm. He has a normal capillary refill. You see that he is breathing spontaneously and, as just noted, is tachypneic and has an increased work of breathing. (Remember that increased work of breathing is synonymous with difficulty ventilating). He is awake and only very mildly ill appearing, but you're able to make the child smile without too much difficulty. You then place a pulse oximeter on the patient's finger and see that his O_2 saturation in ambient air is 98%. Now let's ask ourselves these same emergency questions; after all, few emergencies develop in medicine without any warning signs!

1. *Is air getting to where it needs to be?*
   It is, but with difficulty. You appreciate stridor that occurs only during inspiration, which is reassuring (remember from the previous chapter how expiratory stridor is a medical emergency); however, his increased work of breathing and tachypnea indicate that he is having problems ventilating.
2. *Is the oxygen getting appropriately from the lungs into the blood?*
   He is oxygenating appropriately because his color is good and his pulse oximeter indicates a normal O_2 saturation, but that is likely in part due to his compensatory increase in respiratory effort.
3. *Does the patient have an adequate capacity for transport delivery of O_2 assuming we can get more oxygen into the blood?*

Well, his color is nice and pink. He's got good pulses and good capillary refill. While many children in this age range do suffer from iron-deficiency anemia, it's unlikely to be related to this acute problem, especially given that this patient has stridor. It's safe to say that he has an adequate O_2 delivery capacity.

Distilled down, what's the patient's basic problem, and what are you going to do to stabilize it?

As we've discussed in Chapter 8, stridor is a sign of upper airway obstruction. Given the patient's age and recent signs of a viral illness, an initial assessment of croup causing upper airway obstruction would be appropriate. Most likely a child with increased work of breathing and tachypnea at rest would have biphasic stridor if agitated and crying, or if he began to exert himself physically. Given that this child has a fairly moderate degree of obstruction, acute treatment of nebulized racemic epinephrine and oral steroids to help reduce the swelling would be appropriate. It also would likely be prudent to admit this patient for overnight observation in a hospital. Recall that children have small airways to begin with, and very small changes in worsening of swelling can cause *significant* clinical impact. Given that he already has moderate impairment of ventilation, overnight observation is probably a good idea.

CASE #3

You are a first-year medical resident on your first overnight shift and get called urgently to see a 75-year-old woman who is post-op day 3 from hip surgery. Per the nurse taking care of her, she noticed that the woman began suddenly complaining of shortness of breath and a bit of chest pain. Nursing noted that her pulse oximeter was alarming at 89%. She started her on nasal cannula but the O_2 saturation improved minimally. Upon your arrival you find the woman looking a bit anxious, with an increased work of breathing and a respiratory rate in the high 20 s. The nurse notes that ever since the surgery she has been refusing to get out of bed due to pain in her legs and hip.

What do you do next?

The first thing you would do, much as in the other cases, is a very rapid cardiopulmonary physical examination. You examine the patient and find her to have some accessory respiratory muscle use and her respiratory rate has increased a little into the low 30 s from the low 20 s where she had been the previous day. The nasal cannula is in place and the flow meter indicates she is on 2 L/min of oxygen. Her oral mucous membranes are pink and her pulse oximeter reads 90%. She has palpable pulses, and you estimate her heart rate to be in the 120 s. Her blood pressure is 115/75. Her capillary refill is 2 seconds. You listen to her lungs and hear good air entry bilaterally and no stridor, wheezes, crackles, or rhonchi.

So let's go through our questions:

1. *Is air getting to where it needs to be?*
   At this point it's a little unclear. Your initial suspicion should probably be "yes" given the clear lung sounds and good air entry and the clinical presentation, but some additional questions and physical examination should be performed, as well as some diagnostic testing. You inquire about any medications and you discover that the surgical resident forgot to put this patient on deep venous thromboembolism (DVT) prophylaxis with heparin. Usually patients who are at risk for blood clots (like those older patients who will be bedridden for a time after a lengthy surgery) are typically started on a low dose of a clot-prevention ("blood thinning") medicine called heparin. The patient mentions that she noticed her left leg was swollen, and that this started about 24 hours ago. You're now really concerned that the combination of surgery, immobility, and lack of DVT prophylaxis led this patient to develop a clot in her leg that has now travelled to the lungs. You check a STAT portable chest X-ray, which shows no signs of pneumonia and a normal heart size. You check an EKG, and it shows no evidence of an acute heart attack, and confirms the tachycardia. After calling for help from your upper-level resident and under the resident's supervision you perform an arterial puncture in the patient's left radial artery and obtain an arterial blood gas. It reads as follows: pH 7.48, $PaCO_2$ 28 mmHg, PaO_2 65 mmHg. The hemoglobin is 12 g/dL. The pH is slightly high so the patient is suffering from an alkalemia. The $PaCO_2$ is low, so that would most likely indicate a respiratory alkalosis (remember from Chapter 7, that a respiratory alkalosis means exhaling too much CO_2 thereby decreasing the amount of acid in the blood). Knowing that the patient is able to breathe off CO_2 suggests that she's ventilating

fairly well. In fact, she's hyperventilating. So, she's hyperventilating but the PaO_2 is a bit low. What's going on?

2. *Is the oxygen getting appropriately from the lungs into the blood?*
   With a PaO_2 of only 65 mmHg, you could just ballpark guess that the answer is no. Having recently read *Back to Basics in Physiology: O_2 and CO_2 in the Respiratory and Cardiovascular Systems*, however, you remember that you can evaluate if O_2 is diffusing appropriately between the lungs and the blood using the A−a difference. And, you can actually calculate the A−a difference from the information you have of this patient. The only value we have to figure out at this point is the patient's FiO_2 at the level of the alveoli. Normally, we'd say it's 21%; however, this sample was obtained from a patient who was on 2 L nasal cannula. A fair estimation would be that this patient's alveolar FiO_2 is closer to $\sim 28\%$ (24% FiO_2 with 1 L/min nasal cannula and $\sim 4\%$ FiO_2 increase for every 1 L/min increase). Given this, we can plug these numbers into our equation to solve for P_AO_2.

$$P_AO_2 = FiO_2(P_{ATM} - P_{H_2O}) - \frac{P_aCO_2}{RQ}$$

So,

$$P_AO_2 = 0.28(760 \text{ mmHg} - 47 \text{ mmHg}) - (28/0.8)$$
$$= \sim 200 - 35 = 165 \text{ mmHg}.$$

$$A - a \text{ difference} = P_AO_2 - PaO_2$$
$$= 165 \text{ mmHg} - 65 \text{ mmHg} = 100 \text{ mmHg}$$

It's safe to say that's more than a 10 mmHg difference! So, you're pretty confident the air is getting to the alveolus, but clearly there is a problem with the oxygen making its way into the blood.

3. *Does she have an adequate capacity for transport and delivery of O_2 assuming we can get more oxygen into the blood?*
   The hemoglobin concentration of 12 g/dL is a reasonably normal number for someone who just underwent a major surgery. This patient has also never had problems like this in the past (e.g., problems with her red blood cells). She has a normal blood pressure and an elevated heart rate. The patient is still pink and still has good pulses and good signs of perfusion to the periphery of her body. Thus, her delivery of oxygen should be good if we were able to bring her O_2 sats up. The one thing to consider in this regard is the cardiac output. As we stated previously delivery of O_2 is

contingent on having a good cardiac output. For now, this patient's heart is a little fast, but she has good pulses, good blood pressure, and she's pink, which is a good sign. However, keep the cardiac output in the back of your mind.

Distilled down, what's the patient's basic problem, and what are you going to do to stabilize it?

If you haven't figured it out by now, this patient is suffering from an acute pulmonary embolism (i.e., an acute blood clot in her lungs). Most likely, the clot originated in the veins in her hip or lower leg and eventually broke off, traveling with the veins to the right side of the heart and up into the lungs.

We've established that the problem is not one of ventilation, but of V/Q mismatch. Specifically, in the case of pulmonary embolism, it's easiest to think of a very severe blood clot. With a very severe pulmonary embolism, the embolic event is so large that Q (perfusion) to the affected segments is essentially 0. The blood flow from the right side of the heart divides into the right and left pulmonary arteries. These are big vessels and occasionally, an embolism occurs right where this branching takes place (called a "saddle embolus") and it completely blocks almost all blood flow to the lungs. Because the heart relies on blood returning from the lungs to push out to the rest of the body, if blood flow to the lungs is zero, then eventually so will be the blood flow to the rest of the body. This would cause a build up of blood in the right ventricle, which would lead to acute right ventricular failure, and a massive drop in blood pressure. This patient has a normal blood pressure and, although hypoxemic, is not experiencing profound hypoxemia because her PE is milder.

With a mild PE, the clot is smaller and thus occludes a smaller number of pulmonary capillaries. If blood flow in the affected pulmonary capillaries is minimal or even absent, the blood must go somewhere else to make its way through the pulmonary circulation. This means that a relatively larger amount of blood will then flow through the areas *not* affected by the PE. This larger amount of blood flow can be too much relative to the normal amount of ventilation received in those areas. In other words, in the affected areas the V is normal and the Q is zero. In other areas, the V is normal, and but the Q is too high. Giving this patient oxygen would increase P_AO_2 in all the alveoli.

While this effect would do us no good for the areas that receive no flow, it would do us a world of good for the areas that receive too much blood flow, since they will now carry more oxygen than they otherwise would be able to. Thus, our first decision should be to provide more oxygen to this patient. We could give her a higher amount of nasal cannula flow, however, a non-rebreather mask, which provides more O_2, would also be appropriate in the short-term acute management. (Keep in mind that if the PE is more severe, the positive effect of O_2 might be lost. This is due to a massive excess blood flow to the unaffected segments of the lung to the point where even massive amounts of O_2 can't fully oxygenate the blood that is going through.)

In order to confirm that this is indeed a PE, you could potentially order a CT scan (preferably with IV contrast) that will show an area of the pulmonary circulation where no blood flow is occurring. The second step after confirming this is indeed a pulmonary embolism would be to stabilize the clot by starting the patient on high-dose anticlot medication such as a continuous heparin drip. Although bleeding is a risk for this patient given recent surgery, we have to stabilize this particular patient or she will die from continued pulmonary embolisms! Any problems with her surgical wound can be dealt with should they arise, but not giving someone a potentially life-saving drug to avoid a *potential* problem is likely not the smartest move.

Word of warning, to anticoagulate postoperative patients is by no means an easy decision. This is something that must be evaluated on a case-by-case basis, and an appropriate risk—benefit analysis has to be undertaken by the medical team taking care of every patient.

CASE #4

A 65-year-old previously healthy female presents with a few days' history of fever and cough. The cough has steadily gotten worse, and she has begun feeling progressively more short of the breath over the past couple of days. She is previously healthy and denies any history of medical problems or any medications.

What do you do next?

Well, first thing is always to get the most important information as quickly as possible. There's no quicker or more important test than a physical exam! And getting a brief cardiopulmonary exam focusing on vital signs reveals the most important information. Depending on the findings and the severity of those findings, you can determine the next steps. So you evaluate the patient and see that she is somewhat ill appearing and presently febrile to 103°F (39°C). She is tachypneic with a respiratory rate of 28 and has some increased work of breathing. Her heart rate is 114 with good pulses and normal capillary refill. Mucous membranes are pink in color. On auscultation you appreciate crackles in the left lower lung fields. No wheezing or rhonchi are appreciated. Air entry is diminished in the affected left lung fields, but it is normal on the right. Normal heart sounds are heard. You hook her up to a pulse oximeter and find that her O_2 saturation is 87% without any supplemental oxygen.

1. *Is air getting to where it needs to be?*
   In short, no. You hear diminished breath sounds in the left lower lung fields and sounds of alveoli snapping open; implying that they are inappropriately closed given this increased work of breathing. Increase in minute ventilation with tachypnea and labored breathing is occurring at present to try and make up for this.
2. *Is the oxygen getting appropriately from the lungs into the blood?*
   It would appear some oxygen is, especially on the right. The O_2 saturation is clearly abnormal, but there is still oxygenation taking place. Given our exam findings, we're more and more confident that this hypoxemia is due to a "below the clavicle" problem. V/Q mismatching is likely the culprit. There is poor ventilation on the left, and here O_2 is not appropriately making its way into those particular capillaries. Decrease in oxygenation of that portion of the lung capillaries causes them to constrict, pushing more and more blood flow to the good areas of the lung. However, there comes a point where the lung is unable to supply all that blood with enough ventilated alveoli to get sufficient oxygenation into the blood. This is especially true in times of increased metabolic demand for oxygen, such as with exertion or fever.
3. *Does she have an adequate capacity for transport and delivery of O_2 assuming we can get more oxygen into the blood?*
   Yes, as her heart is presently working strong. Given her pink color and lack of prior medical problems means she should have a sufficient O_2 carrying capacity, she just needs more oxygen to make its way into the blood.

Distilled down, what's the patient's basic problem, and what are you going to do to stabilize it?

Clinically, this presentation is most likely a bacterial pneumonia. You would always want to optimize oxygenation and ventilation prior to doing any additional testing, and so it would be reasonable to start her on a nasal cannula or non-rebreather mask. Starting antibiotics would also be warranted, and a confirmatory chest X-ray would not be unreasonable either, but only *after* you have succeeded in stabilizing this patient. So, you start this patient on a non-rebreather mask, and her oxygen saturations improve to the mid 90 s, and you send her for a chest X-ray. The chest X-ray shows a large left pneumonia covering much of the left lung at which point you decide to start antibiotics. The patient begins to complain of more shortness of breath and you find that her O_2 saturations have again dipped into the high 80 s. At this point it would probably be appropriate to try the patient on positive pressure to help stent open the underventilated portions of the left lung to try and maximize surface area for gas exchange. After starting the patient on BiPAP, you find that her saturations again improve, as does her work of breathing. You admit this patient to the ICU for additional monitoring/treatment.

CASE #5

A 14-year-old male presents to your office with a 2-day history of cough and an interval development of difficulty breathing that started acutely today. This patient has a history of asthma as well as a history of poor compliance with his medications. He states that he ran out of his inhaler a few days ago.

What do you do next?

Again, physical exam. You look at the patient and find that he has a respiratory rate in the low 20 s (slightly elevated). He appears a bit apprehensive. He has some accessory muscle use (which means contractions of the diaphragm are no longer sufficient to move air in and out of the lungs). His color is pink. Heart rate is 110 with normal heart sounds. On auscultation there is limited air entry bilaterally. You appreciate some faint wheezing, but overall air entry is difficulty to hear. You hook him up to a pulse oximeter and see that his O_2 saturation is 91%.

1. *Is air getting to where it needs to be?*
   No, which appears to be fairly evident in this case as you can't appreciate good air entry anywhere on auscultation, and this patient has signs of increased work of breathing.
2. *Is the oxygen getting appropriately from the lungs into the blood?*
   Based on his color and pulse oximetry, yes. But there is some difficulty, which appears largely due to impaired ventilation at this time.
3. *Does he have an adequate capacity for transport and delivery of O_2 assuming we can get more oxygen into the blood?*
   It would seem reasonable to assume so for the same reasons mentioned in the other cases. Pink color, no signs of pallor, no history of red blood cell problems.

   Distilled down, what's the patient's basic problem, and what are you going to do to stabilize it?

   This patient is presenting with an acute asthma attack. He will likely need oxygen, but at this moment we also need to acutely improve his ventilation. The two mainstays of therapy are inhaled beta agonists such as albuterol and oral or IV steroids. This is because in asthma the ventilatory problems are two-fold. On one hand, you have bronchospasm that is caused by the muscles within the lower airways tightening and causing narrowing of the airway. This is relieved by albuterol. However, a more worrisome component of asthma is actually the inflammation that accompanies it. Depending on how severe and poorly controlled this patient's asthma is, he can have varying degrees of airway inflammation. Not only is the airway narrowed due to constriction of the muscles, but it's also narrower because the walls of the airway themselves become swollen. There is increased mucous production as well as cellular debris, which gets lodged down into the lower airways and causes easy plugging above the alveoli. The lack of breath sounds, as well as hypoxemia (remember the low O_2 saturation), is suggestive of a more severe asthma attack (e.g., as opposed to a patient who has good air entry but is wheezing at the end of expiration). So we give this patient several inhaled treatments of albuterol as well as oral steroids. Shortly after the albuterol treatments, however, we see that his O_2 saturations have actually worsened to 88%. On auscultation he is now moving air better and you can appreciate diffuse inspiratory and expiratory wheezing. The patient states he feels like he can breathe a little better.

Why did this happen? What are you going to do now?

Although the drop in O_2 saturations could have been due to worsening of his acute asthma attack, it seems unlikely given that other indicators have improved (namely his subjective report of dyspnea and better air entry on auscultation). In general in medicine, it's often better to treat the patient before the number, but in this case his drop in oxygen levels can be relatively easily explained. He is still having an asthma attack. The inflammation is still present as is some component of bronchospasm. However, the albuterol has helped improve the bronchospasm. Recalling our normal V/Q mechanisms, it would make sense that the initial drop in O_2 was due to poor ventilation. The body compensates for that by constricting blood flow to the poorly ventilated areas. If we've improved ventilation with albuterol, why has the hypoxemia worsened slightly? Again, this is a "below the clavicle" problem, so would you be surprised to learn that the answer is again V/Q mismatch? Albuterol has a side effect of dilating not just muscles within the airways, but also within the small arteries (arterioles) of the lung. This causes increased blood flow throughout the entire lung, including to the areas that are badly inflamed and still not ventilating well. This is a very common finding with albuterol to see someone's O_2 saturation drop slightly. You combat this by simply increasing FiO_2 concentration by giving the patient supplemental oxygen, typically via simple facemask or nasal cannula. Be aware that this increase in blood flow due to albuterol should cause only a *small* drop in O_2 saturation. Assuming normal O_2 carrying capacity and no other problems, profound hypoxemia in the setting of asthma would suggest either a very severe asthma attack or some other problem.

CASE #6

You're working your first shift ever as a resident in the Emergency room and a 22-year-old man presents with agitation, tachycardia, and tachypnea. He was found on the ground with a gunshot wound to the leg near the hospital and brought in by ambulance.

What do you do next?

Physical exam. You evaluate this patient and find him to be a bit confused. He has some pallor. His respiratory rate is in the mid 20 s and his heart rate is 140. His blood pressure is 95/80. His capillary refill is delayed (which is a sign of poor blood flow) and his pulses are palpable, but a bit weak. His extremities are cool to the touch. You then hook him up to a pulse oximeter and see that his O_2 saturations are 99%. You listen to his lungs and hear good air entry without abnormal sounds. Normal heart sounds are heard. No other signs of trauma appreciated elsewhere.

1. *Is air getting to where it needs to be?*
   Yes. Lung sounds are good. There was no evidence to suggest problems within the lungs prior to this gunshot wound. Oxygen saturations are appropriate.
2. *Is the oxygen getting appropriately from the lungs into the blood?*
   Given the normal oxygen saturations, yes.
3. *Does he have an adequate capacity for transport and delivery of O_2 assuming we can get more oxygen into the blood?*
   NO! Remember, delivery of O_2 of determined by hemoglobin content within the blood and cardiac output. Just because 99% of the hemoglobin is saturated, based on clinical findings this patient has acutely lost between a quarter to a third of his entire circulating blood volume! So, sure, the hemoglobin that's there is well saturated, but there are a lot fewer red blood cells to carry the load. If you think of it as box cars on a train, each car may be fully loaded, but if a third of the train gets detached and doesn't arrive, that's a lot of missed cargo! The key point to understand here is that a normal O_2 saturation is expected in this case, and it is *not* reassuring that this patient is doing OK.

Distilled down, what's the patient's basic problem, and what are you going to do to stabilize it?

What's wrong with the patient should be fairly obvious. Acute blood loss! He does not have a ventilation problem, he has a hemoglobin problem. And he's rapidly developing a problem with delivery due to inadequate cardiac output should blood loss persist, because there won't be enough blood to maintain blood pressure. In order to stabilize this patient, O_2 would be helpful, as it would increase the amount of O_2 dissolved in blood but not bound to hemoglobin. However, most important would be improving blood volume and red blood cell content. So acutely the patient would need fluid resuscitation and red blood cell transfusion. This of course would be followed closely by the need for a surgeon to help stop further bleeding by stopping further hemorrhage.

Respiratory Devices

Oxygen itself was discovered at the end of the 1700 s. Scientists discovered that by heating mercuric oxide (don't worry, there won't be quiz on this, it's just for fun), a candle flame burned brighter. At the time they believed by heating the mercuric oxide, they were removing impurities from the air. Upon breathing this "purer air," which we now know was concentrated oxygen, Joseph Priestley, the first person to publish something on the subject, had this to say about the experience:

> The feeling of it to my lungs was not sensibly different from that of common air; but I fancied that my breast felt peculiarly light and easy for some time afterwards. Who can tell but that, in time, this pure air may become a fashionable article in luxury. Hitherto only two mice and myself have had the privilege of breathing it.
> **Joseph Priestley, Experiments and Observations on Different Kinds of Air**

While it may not be a luxury, supplemental O_2 has revolutionized medicine and prevented untold morbidity and mortality (OK, maybe they use it in casinos and at oxygen bar parties, so maybe Joseph Priestley wasn't all that wrong, but we digress).

We've previously established that ventilation is required to deliver O_2 to the alveoli, and that oxygenation of the blood results when there is a sufficient partial pressure gradient of O_2 within the alveoli. We've also addressed some conditions in which this may be problematic, such as in the case of respiratory disease. Thankfully, in a beautiful symphony of medical and engineering talent, the advent of respiratory medical devices was born.

Supplemental O_2 is no longer obtained from heating mercuric oxide; rather industrial production is most commonly performed by cooling air to less than $-320°F$, making it a liquid. Because the different components of air have differently boiling points, they are able to raise the temperature slowly and boil off the N_2 and retain the O_2 (via distillation). It is then commonly cooled once again and transported and stored as a liquid. This makes sense when we remember to think of these matter states as particles, with gas particles taking up lots of space and easily compressible, and liquids being much more compressed. As such, for every liter of liquid oxygen, when warmed up a bit it becomes 840 L of gaseous O_2! That's pretty economical in terms of space-saving, no?

FIRST, A BRIEF PRIMER ON GETTING TOO MUCH O_2

Because this topic is very involved and the minutiae largely beyond the scope of this book, we can simply state for the purposes of supplemental oxygen as a medical therapy that too much of it can be a bad thing. Although O_2 made Mr. Priestley's breast feel peculiarly light and easy, too much of it could have had the opposite effect! We know all too well at this point that oxygen is utilized in metabolism to create more efficient ATP production. What you may *not* know, however, is that during the normal metabolic processes something called reactive oxygen species (ROS) are created. Like most things in life, metabolism is not always perfect. This species, such as oxygen free radicals ($O_2^{\bullet}$), hydrogen peroxide (HOOH), superoxide anion (O_2^-), and others can all zing around and injure various parts of the body. Most notably, these radicals can damage DNA, as well as proteins and fats that are especially prominent in the brain and lungs. Normally, there are incredibly efficient enzymes that quickly fix these ROS and convert them back into things like O_2 and H_2O; however, these enzymes are limited in number. By giving a patient too much supplemental O_2, you end up increasing O_2 concentrations throughout the entire body, thus exponentially increasing the amount of ROS production. This production can overwhelm these protective enzymes and end up causing damage to the lungs in the form of clinical conditions with names such as diffuse alveolar injury and acute respiratory distress syndrome.

OXYGEN DELIVERY DEVICES

Now that that's out of the way, let's look at how we (responsibly) deliver this O_2! In this appendix we will look at the various O_2 delivery forms that you will commonly see employed in a hospital for patients with acute respiratory problems. As we go through these devices, keep in mind that the best way to clinically integrate them is to ask a respiratory therapist at your hospital to walk you through these devices, so you can see, touch, and understand them a whole lot better.

Nasal Cannulae

Nasal cannulae (plural for cannula as there are two of them, one for each nasal opening) is a comfortable, easy method for oxygen delivery when a lesser amount of alveolar FiO_2 is needed. They are composed of plastic tubing that has two prongs that are placed one in each nostril. Standard nasal cannulae generally allow for a flow of 0.25 L/min to 6 L/min, however, in very small infants even lower flow rates can be seen. Depending on flow rate, actual alveolar FiO_2 delivery can be seen ranging between 25 and 50%. It is generally accepted that there is a 4% increase in FiO_2 delivery for every L/min increase in flow. The actual amounts of FiO_2 delivery in practice are highly variable for reasons discussed previously. If you require higher amounts of FiO_2 because you are acutely ill, chances are you will also be breathing faster and more profoundly. This results in higher alveolar minute ventilation that in more serious cases may be higher than oxygen delivery.

Some limitations of these basic nasal cannulae are that they are small and also allow for entrainment of ambient air from the environment. This entrainment occurs around the cannulae themselves if the patient is breathing especially quickly or profoundly through the nose, but especially if the patient is breathing through the mouth.

Simple Face Mask

A simple face mask is designed to fit over both the nose and the mouth. These masks offer FiO_2 delivery between 35 and 55%. In its simplest form, this type of mask has two holes in it to allow the patient to both inhale and exhale ambient air, which limits the total amount of FiO_2. Some masks have one-way valves, which limit the inhalation of ambient air, but do not prevent the exhalation of a patient's breath. This ambient air "dilution" is offset by increasing the flow through the oxygen port, often at least 8 L/min or greater. These masks are also convenient for delivery of certain aerosolized medicines (e.g., albuterol in patients with an acute asthma attack).

Limitations are similar to other lower-flow devices in that there is only so much FiO_2 that can be delivered. Also, a minimum set flow rate (usually at least 4–6 L/min depending on mask type and patient size) needs to be set to ensure the patient is also not rebreathing his or her own exhaled CO_2. Because a guaranteed minimum is required, this may end up causing more oxygenation than desired.

Venturi Mask

A Venturi mask is similar to a simple face mask but has the added benefit of being able to set a fixed amount of FiO_2 to be delivered. This is accomplished by switching out the simple oxygen port in the simple mask, and replacing it with a device that delivers a set amount of room air along with the oxygen, thereby allowing the provider to set the desired FiO_2. This is particularly helpful in settings where you only have means of delivering oxygen and not room air. However, in most hospital settings, they can employ something called an oxygen blender. Rooms are equipped with both a nozzle for both 100% oxygen and a nozzle for ambient air. A blender is a device that connects both nozzles and allows "blending" gases from the two ports. This way, you can deliver less oxygen right from the wall, and it can be used with almost all the oxygen delivery devices!

Non-rebreather Mask

A non-rebreather mask is similar to a Venturi mask or a simple mask, but it is designed to limit the amount of entrained ambient air. Because the earlier devices can only deliver a set amount of oxygen that can easily be diluted by entraining ambient air, the non-rebreather mask corrects these limitations. As we said, a limitation of nasal cannulae is entraining room air through the mouth or around the nasal cannulae through the nose. The limitation of facemasks is entraining room air through the ports or by rebreathing exhaled air. The non-rebreather mask fixes these problems by having two one-way valves that allow exhalation of air but no inhalation of ambient air. Rebreathing exhaled air is prevented by giving a large amount of oxygen flow to help keep air within the mask constantly circulated with fresh oxygen. However, should a patient need larger breaths, a large reservoir bag is added near the site of the oxygen port as well that is constantly being filled with supplemental oxygen. In this way, you can deliver the highest amount of FiO_2 available through simpler means. It is theoretically just a little less than 100% FiO_2.

The problem is, however, what happens if the patient were to become disconnected from the oxygen? If the patient was unable to take the mask off him- or herself, he or she could suffocate! It's much like putting a tight-fitting paper bag around your face! Therefore, legally it is required that one of the one-way ports be removed or left open to allow entrainment of ambient air as a safety mechanism should the patient become disconnected from the oxygen. The non-rebreather mask is one of the most common oxygen delivery systems utilized in emergency rooms throughout the country! This is because in an acute setting, you're not worried about delivering too much oxygen. You're just worried about getting the patient oxygen, and fast!

An important limitation of the non-rebreather mask, in addition to possibly providing *too much* oxygen, is that it does nothing to help with ventilation! This is true of *all* the devices listed in this section! Thus, you can have 100% FiO_2 delivered all day long, but if it isn't getting down to the alveoli, it doesn't matter! Thus, we will now look at some of the simpler ventilation devices, more commonly known as positive pressure ventilation (PPV).

A BRIEF PRIMER ON POSITIVE PRESSURE VENTILATION

As we've discussed at some length, if you're not ventilating, you're not going to oxygenate. In cases where there is a problem with mild-to-moderate V/Q mismatching and such, increasing the concentration of oxygen delivered to the properly functioning alveoli (hyperoxygenating these areas) is often sufficient to result in overall adequate amounts of blood oxygen content and subsequent oxygen delivery to tissues. However, in other circumstances where ventilation is more seriously impaired, simply increasing oxygen percentage in the inspired air is *not* sufficient. If ventilation is the primary problem, then the only way to correct the subsequent inadequate blood gas exchange is to fix the underlying problem: ventilation!

Early attempts to fix this problem were with something called an iron lung machine. In a way, these were both more complicated but also simpler devices in that they attempted to aid the body to breathe in a way that was more natural. The machine would encapsulate a person from the neck down in a box that was sealed around the patient's neck. Essentially with the ventilator we have another balloon-in-a-bottle problem. The machine would act as a pump and a vacuum, both sucking air out from around the patient's body and then pumping it back in. This would effectively change the *atmospheric* pressure rather than the pressure within the patient's body. In the early twentieth century, polio still ravaged many pediatric patients, and in severe forms it would cause paralysis and an inability to breathe unassisted. These iron lung machines kept many people from dying.

However, as with many of the world's technological advancements, technology utilized during war (World War II) helped fuel technological advancement in other areas. Research into how to better oxygenate aircraft pilots operating at high altitude was applied to newer positive pressure ventilators. If we go back to our balloon-in-a-bottle example we can analyze PPV in its simplest form. In a way, although not physiologic, this type of breathing makes a lot more sense, as frequently problems ventilating are not simply due to inadequate respiratory effort! As we discussed in the last chapter, oftentimes problem ventilating is accompanied by vigorous respiratory effort. The advantage of PPV is that you can help overcome some of the limitations of negative pressure ventilation by simply forcing open areas of the lung that would otherwise be closed due to mucus, debris, swelling, and so on. The question, however, is how do we deliver this positive pressure?

Noninvasive Positive Pressure Ventilation
High-Flow Nasal Cannulae
Perhaps the simplest (and least well understood) respiratory device that provides a component of positive pressure is the high-flow nasal cannula (HFNC). Essentially, these are better fitting nasal cannulae. Their tube diameters are larger to accommodate a higher flow, as well having the added benefit of resting more snugly in the nostrils. Although there is still air leak, it is typically less than seen with standard nasal cannulae. The tubing that connects the cannulae is large bore to prevent resistance within the system. It is also humidified and warmed via an external machine. With high oxygen flow rates to a maximum of 40 L/min for adults, you could see why this humidification and warming of oxygen is so important! At those volumes, you'd dry out your nose and airway as well as cause significant heat loss!

The limitations of HFNC are significant. For one, though larger bore, the nasal cannulae do not necessarily form an airtight sea. Thus air dilution and pressure loss can occur both around the cannulae and through the mouth. As such there are unpredictable amounts of positive pressure.

At times, pressure can be rather high, and other times less so. Also, for these same reasons, oxygen delivery is not as high as that of a non-rebreather mask. Furthermore, if your problem with ventilation is due to inadequate respiratory effort, then HFNC does little to help with this problem.

CPAP

CPAP stands for continuous positive airway pressure. It is most commonly delivered via a form-fitting mask that is tightly but comfortably adherent over the nose and mouth. It is kept tight fitting thanks to elastic straps that wrap around the back of the head. Because there is a seal overlying around both the nose and mouth, pressure can be set via a machine that will deliver flow to meet a set pressure. This pressure is continuous and similar to something called PEEP, or positive-end expiratory pressure, settings you might see on a ventilator. CPAP's goal is to make it stent open the airways, aid in lung recruitment, and prevent alveoli from closing. It can also help stent open the upper airways, such as the oropharynx, and is thus particularly useful in patients with obstructive sleep apnea.

One of the limitations of CPAP is that it only helps prevent the collapse of the airway. It does not, however, aid in ventilatory effort on inspiration aside from this. While it is true breathing is easier if the alveoli remain partially open (again, think of blowing up a balloon), CPAP does very little else to aid in inspiratory effort. Thus, the onus is still on the individual to inhale effectively.

BiPAP

BiPAP, or bi-level positive airway pressure, helps to make up for the limitations of CPAP. BiPAP allows for two different pressures to be set, one inspiratory (IPAP) and one expiratory (EPAP). The ventilators that drive BiPAP are able to give these breaths spontaneously; that is, when the machine senses a drop in pressure due to a patient's inspiratory effort, it will trigger oxygen flow to be delivered until a desired inspiratory pressure is reached. These machines are also capable of giving inspiratory positive pressure at a set backup rate as well, in cases where patients have neurological conditions that prevent them from initiating breaths consistently. Most commonly, however, a combination of the two modes is the preferred method of delivery. The mask and ventilator system, similar to CPAP, has features within the system to allow for expiratory air leak so as to avoid CO_2 retention.

Both CPAP and BiPAP can also be delivered via nasal prongs that are much larger and more firmly occlude the nasal passages.

BiPAP has limitations that are also shared with CPAP. Both means of noninvasive PPV are delivered through similar facemasks. As such, they can cause skin breakdown with prolonged use. Also, because the ventilation is being provided to oro- and nasopharynx, there is risk for the air to enter the stomach as well. Patients are at risk for nausea and vomiting as a result. In younger patients or older patients who are disabled and unable to take off the mask in the event they feel they may throw up, this poses a serious potential complication! Throwing up in your BiPAP mask is *not* a good thing to do!

Not only is air getting into the stomach a problem, but a problem shared by all modes of noninvasive PPV is that it's simply not as effective as delivering the oxygen *directly* into the lungs. For this, medicine created a solution called an endotracheal tube.

Endotracheal Intubation and the Conventional Mechanical Ventilator

The *invasive* means of PPV also happens to be the simplest to understand (and the hardest to master). If you needed to ventilate the lungs by delivering positive pressure, what better way than putting a tube directly into the trachea! By developing tubes that had a soft pressure cuff on the outside that you can inflate with air, you could effectively create a delivery mechanism that is the most direct and effective means of ventilation possible. These tubes can be inserted either through the nose, the mouth, or by cutting into the neck and inserting them directly through the cricothyroid cartilage (talk about invasive!). By doing so with an inflated cuff that is gently inflated until the airway around the tube is occluded, you've effectively created a simple balloon-in-the-bottle scenario yet again. With this system, there are fewer variables to work with since you're working with the lungs right from the source. That said, you're also making it very difficult for the patient to breathe on his or her own, and thus there is less room for error and more need for closer monitoring. With the ventilator, you can control both inspiratory and expiratory parameters and you can do so based on delivering set tidal volumes, set pressures, or a combination of the two. As interesting as ventilatory and critical care medicine is, it is a bit beyond the scope of this text, since this is something that is really principally managed by specialists, namely critical care physicians or pulmonologists.

BIBLIOGRAPHY

GENERAL REFERENCES

Arroyo, J.P., Schweickert, A., 2013. Back to Basics in Physiology: Fluids in the Cardiovascular and Renal Systems, 1st ed. AP/Elsevier, Philadelphia, PA.

Boron, W.F., Boulpaep, E.L., 2009. Medical Physiology: A Cellular and Molecular Approach, 2nd ed. Saunders/Elsevier, Philadelphia, PA.

Hall, J.E., Guyton, A.C., 2011. Guyton and Hall Textbook of Medical Physiology, 12th ed. Saunders/Elsevier, Philadelphia, PA.

Marino, P.L., 2013. The ICU Book, 4th ed. Lippincott Williams & Wilkins, Philadelphia, PA.

West, J.B., 2008. Respiratory Physiology: The Essentials, 8th ed. Lippincott Williams & Wilkins, Philadelphia, PA.

SPECIFIC REFERENCES

Ballester, E., Reyes, A., Roca, J., Guitart, R., Wagner, P.D., Rodriguez-Roisin, R., 1989. Ventilation-perfusion mismatching in acute severe asthma: effects of salbutamol and 100% oxygen. Thorax 44 (4), 258–267.

Blumenthal, I., 2001. Carbon monoxide poisoning. J. R. Soc. Med. 94 (6), 270–272.

Cheifetz, I.M., 2011. Pediatric acute respiratory distress syndrome. Respir. Care 56 (10), 1589–1599. Available from: http://dx.doi.org/10.4187/respcare.01515.

Chow, D.C., Wenning, L.A., Miller, W.M., Papoutsakis, E.T., 2001. Modeling pO(2) distributions in the bone marrow hematopoietic compartment. II. Modified Kroghian models. Biophys. J. 81 (2), 685–696. Available from: http://dx.doi.org/10.1016/S0006-3495(01)75733-5.

Efthimiou, J., Mounsey, P.J., Benson, D.N., Madgwick, R., Coles, S.J., Benson, M.K., 1992. Effect of carbohydrate rich versus fat rich loads on gas exchange and walking performance in patients with chronic obstructive lung disease. Thorax 47 (6), 451–456.

Frayn, K.N., 1983. Calculation of substrate oxidation rates in vivo from gaseous exchange. J. Appl. Physiol. Respir. Environ. Exerc. Physiol. 55 (2), 628–634.

Glenny, R.W., 2008. Teaching ventilation/perfusion relationships in the lung. Adv. Physiol. Educ. 32 (3), 192–195. Available from: http://dx.doi.org/10.1152/advan.90147.2008.

Hasleton, P.S., 1972. The internal surface area of the adult human lung. J. Anat. 112 (Pt 3), 391–400.

Inwald, D., Roland, M., Kuitert, L., McKenzie, S.A., Petros, A., 2001. Oxygen treatment for acute severe asthma. BMJ 323 (7304), 98–100.

Johnson, J.D., Theurer, W.M., 2014. A stepwise approach to the interpretation of pulmonary function tests. Am. Fam. Physician. 89 (5), 359–366.

Parameswaran, K., Todd, D.C., Soth, M., 2006. Altered respiratory physiology in obesity. Can. Respir. J. 13 (4), 203–210.

Rodriguez-Roisin, R., 1997. Acute severe asthma: pathophysiology and pathobiology of gas exchange abnormalities. Eur. Respir. J. 10 (6), 1359–1371.

Slutsky, A.S., Ranieri, V.M., 2013. Ventilator-induced lung injury. N. Engl. J. Med. 369 (22), 2126–2136. Available from: http://dx.doi.org/10.1056/NEJMra1208707.

Stamati, K., Mudera, V., Cheema, U., 2011. Evolution of oxygen utilization in multicellular organisms and implications for cell signalling in tissue engineeringJ. Tissue Eng. 2 (1), 2041731411432365. Available from: http://dx.doi.org/10.1177/2041731411432365.

Stather, D.R., Stewart, T.E., 2005. Clinical review: mechanical ventilation in severe asthma. Crit. Care 9 (6), 581–587. Available from: http://dx.doi.org/10.1186/cc3733.

Subramani, S., Kanthakumar, P., Maneksh, D., Sidharthan, A., Rao, S.V., Parasuraman, V., et al., 2011. O2-CO2 diagram as a tool for comprehension of blood gas abnormalities. Adv. Physiol. Educ. 35 (3), 314–320. Available from: http://dx.doi.org/10.1152/advan.00110.2010.

Temple, A.R., 1981. Acute and chronic effects of aspirin toxicity and their treatment. Arch. Intern. Med. 141 (3 Spec No), 364–369.

Wagner, P.D., 2008. Causes of a high physiological dead space in critically ill patients. Crit. Care 12 (3), 148. Available from: http://dx.doi.org/10.1186/cc6888.

Wallace, K.B., Starkov, A.A., 2000. Mitochondrial targets of drug toxicity. Annu. Rev. Pharmacol. Toxicol. 40, 353–388. Available from: http://dx.doi.org/10.1146/annurev.pharmtox.40.1.353.

West, J.B., 2011. Causes of and compensations for hypoxemia and hypercapnia. Compr. Physiol. 1 (3), 1541–1553. Available from: http://dx.doi.org/10.1002/cphy.c091007.

Printed in the United States
By Bookmasters